Immunocomputing
Principles and Applications

Springer
*New York
Berlin
Heidelberg
Hong Kong
London
Milan
Paris
Tokyo*

A.O. Tarakanov   V.A. Skormin
S.P. Sokolova

Immunocomputing

Principles and Applications

With 36 Figures

 Springer

A.O. Tarakanov
St. Petersburg Institute for Informatics
Russian Academy of Sciences
14-line 39
St. Petersburg 199178
Russia
tarakanov@togetherlab.nw.ru

V.A. Skormin
Professor of Electrical Engineering
Watson School, Binghamton University
Binghamton, NY 13902
USA
vskormin@binghamton.edu

S.P. Sokolova
St. Petersburg Institute for Informatics
Russian Academy of Sciences
14-line 39
St. Petersburg 199178
Russia
sokolova_sv@mail.ru

Library of Congress Cataloging-in-Publication Data
Tarakanov, Alexander O.
   Immunocomputing: principles and applications / Alexander O. Tarakanov, Victor A. Skormin, Svetlana P. Sokolova.
     p. cm.
   Includes bibliographical references and index.
   ISBN 0-387-95533-X (alk. paper)
     1. Immunocomputers.   2. Molecular electronics.   I. Skormin, Victor A., 1946–
II. Sokolova, Svetlana P.   III. Title.
QA76.875.T37 2003
621.39′1—dc21   2002044508

ISBN 0-387-95533-X   Printed on acid-free paper.

© 2003 Springer-Verlag New York, Inc.
All rights reserved. This work may not be translated or copied in whole or in part without the written permission of the publisher (Springer-Verlag New York, Inc., 175 Fifth Avenue, New York, NY 10010, USA), except for brief excerpts in connection with reviews or scholarly analysis. Use in connection with any form of information storage and retrieval, electronic adaptation, computer software, or by similar or dissimilar methodology now known or hereafter developed is forbidden. The use in this publication of trade names, trademarks, service marks, and similar terms, even if they are not identified as such, is not to be taken as an expression of opinion as to whether or not they are subject to proprietary rights.

Printed in the United States of America.

9 8 7 6 5 4 3 2 1   SPIN 10883971

www.springer-ny.com

Springer-Verlag   New York Berlin Heidelberg
A member of BertelsmannSpringer Science+Business Media GmbH

Preface

Overview

This book introduces *immunocomputing* (IC) as a new computing approach that replicates the principles of information processing by proteins and immune networks. It establishes a rigorous mathematical basis for IC, consistent with recent findings in immunology, and it presents various applications of IC to specific computationally intensive real-life problems. The hardware implementation aspects of the IC concept in an *immunocomputer* as a new kind of computing medium and its potential connections with modern biological microchips (biochips) and future biomolecular computers (biocomputers) are also discussed.

All biological systems at the cellular and biomolecular levels are sophisticated mechanisms honed to perfection by millions of years of evolution, and their exploration provides inspiration for various novel concepts in science and engineering. Of these systems, however, only two types, the neural system and the immune system of the vertebrates, possess the extraordinary capabilities of "intellectual" information processing, which include memory, the ability to learn, to recognize, and to make decisions with respect to unknown situations.

The potential of the natural neural system as a biological prototype of a computing scheme has already been utilized intensively in computer science through the mathematical and software models of artificial neural networks (ANN) and their hardware implementation in neural computers (see, e.g., Haykin, 1999; Wasserman, 1990).

However, the extraordinary information processing capabilities of the natural immune system have only been recently appreciated (Dasgupta, 1999; Segel and Cohen, 2001; de Castro and Timmis, 2002). The mathematical formalization of these capabilities forms the basis of the new computational approach *immunocomputing* (IC) that is established in this book.

A need for such an approach proceeds from the recent findings of molecular immunology (immune networks theory, role of cytokinins, immune synapse, neuro-immune modulation) and rapidly emerging biochip technology (Cheng and Kricka, 2001). This technology brings together computer chips with biomedical assays and laser-based detectors. Biochips make possible the manipulation of proteins and other biomolecules to perform immunoassays for highly throughput, precise, and economical diagnostics, and even for computer controlled fragments of the natural immune system in the near future.

Audience

The book is intended for experts in computer science, artificial intelligence, and biomolecular computing interested in a thoroughly developed mathematical basis for adopting the "natural" principles of computing that have been honed to perfection by millions of years of evolution, and it is also intended for immunologists seeking to further quantify their field of research, multidisciplinary researchers interested in the mutual enhancement of computer science and immunology, and university students exploring their individual "entry points" to the world of science.

Organization and Features

The book contains seven chapters. The first chapter describes a biological prototype and shows the place of IC among modern information technologies. The second chapter develops a mathematical basis for IC. The third, fourth, and fifth chapters show how important areas of computing could be redefined on this basis. The sixth chapter considers a number of real-life applications of IC. And the final chapter discusses a possible hardware implementation for IC in an immunochip and its followup in a biochip and a future biocomputer.

The first chapter represents an introduction to the book. It provides a sketch of key biomolecular mechanisms of information processing that remain out of the scope of computer science and artificial intelligence (AI). The chapter reveals the potential of IC as a new kind of computing technology.

The second chapter develops a rigorous mathematical basis for IC. The main notions of this basis are formal protein (FP) and formal immune network (FIN). The necessary mathematical notions for these concepts are quaternions and bilinear forms.

The third chapter provides a theoretical framework for pattern recognition in IC. The main notion is binding energy between FPs. The extreme values of the binding energy present a solution to the pattern recognition task.

The fourth chapter provides a theoretical framework for language representation and knowledge-based reasoning by IC. The approach is based on treating linguistic symbols as FPs and linguistic relations as interactions between FPs.

The fifth chapter develops an IC approach to the modeling of natural and technical systems. The examples include, but are not limited to, the modeling of natural proteins, computer networks, and dynamic modeling of the clothing process. These apparently very different problems are computationally intensive and, theoretically, could be formalized as special kinds of FPs and FINs.

A number of IC applications are investigated in the sixth chapter. These are real-life problems chosen from the disparate fields of space navigation, ecology, infection control, and security.

The final chapter suggests a hardware implementation of the IC approach in an immunochip as the core of a future immunocomputer. As a further development of the IC approach, a gradual transformation of a protein biochip to a biomolecular computer is presented.

Acknowledgments

The authors are grateful to the following organizations for their partial support:

- the European Commission (the EU projects IST-2000-26016 "Immunocomputing" and ICA2-CT2000-10048 "The plague of Central Asia – an epidemiological study focusing on space-time dynamics")
- the US Air Force Research Laboratory at Rome NY (the project BASIS – "Biological Approach to System Information Security")
- the European Office of Aerospace Research and Development (the EOARD-ISTC project 2200p "Development of Mathematical Models of Immune Networks Intended for Information Security Assurance)

We would like to thank the following individuals for their help:

- Dr. Andrew Adamatzky at the University of the West of England, Bristol, UK (Section 5.3)
- Sergey Lesovoy and Vladimir Talyzin (Section 5.3.4)
- Dr. Vladimir Kuznetsov (Section 6.2)
- Sergey Kvachev (Section 7.2)
- Dr. Larisa Goncharova and Prof. Tatyana Gupalova (Section 7.3)
- Joseph Giordano, John Graniero, and Dr. Donald J. Nicholson of the AFRL at Rome NY for their ongoing support of the BASIS project

We would also like to express our particular gratitude to Springer-Verlag NY, especially to Wayne Yuhasz, Wayne Wheeler, and Frank Ganz for their interest in this book, invaluable support, and professional patience.

Contents

Preface v

1 Introduction 1

1.1 The Mathematical Basis, Trends and Limitations of Modern Artificial Intelligence.. 1
1.2 Key Mechanisms of Biomolecular Computing 4
   1.2.1 Proteins from the Computing Viewpoint 4
   1.2.2 The Mechanisms of Protein Behavior 5
   1.2.3 The Immune System as a Model for a New Kind of Computing 7
1.3 Artificial Immune Systems.. 8
1.4 Immunocomputing as a New Kind of Computing.................................. 11

2 The Mathematical Basis of Immunocomputing 13

2.1 The Formal Protein.. 13
   2.1.1 The Algebraic Description of the Shape of Natural Proteins 13
   2.1.2 The Mathematical Notion of a Formal Protein................................ 15
2.2 The Interaction Between Formal Proteins 17
   2.2.1 Binding Energy 17
   2.2.2 The Allosteric Effect 18
   2.2.3 Networks of Binding 19
2.3 Formal Immune Networks... 22
   2.3.1 The Formal B-Cell 22
   2.3.2 The Formal T-Cell... 23
   2.3.3 The Mathematical Notion of FIN 23
   2.3.4 The Properties of FINs 27
2.4 Related Topics.. 28

2.4.1 Quaternion Algebra 28
2.4.2 Singular Value Decomposition 31
2.4.3 The SVD of Interval Matrices 36

3 Pattern Recognition 41

3.1 A Mathematical Model of Molecular Recognition 41
    3.1.1 The Representation of a Bilinear Form 41
    3.1.2 Recognition between FPs 44
    3.1.3 Specificity of Recognition 46
3.2 Pattern Recognition by Immunocomputing 50
    3.2.1 The General Task of Pattern Recognition 50
    3.2.2 Supervised Learning 53
    3.2.3 Unsupervised Learning 53
3.3 Main Computing Procedures 54
    3.3.1 Supervised Learning 55
    3.3.2 Unsupervised Learning 57
    3.3.3 Recognition 59

4 Language Representation and Knowledge Based Reasoning 61

4.1 Morphology: Peptide Spectrum of a Word 61
4.2 Syntax: Matrix Eigenlanguages 65
4.3 Knowledge-Based Reasoning 78
4.4 Linguistic Binding 81

5 Modeling of Natural and Technical Systems 83

5.1 Spatial Structures of Native Proteins 83
    5.1.1 Algebraic Description of Secondary Structures 83
    5.1.2 Dependence of the Configuration on the Amino Acid Sequence 89
5.2 Synchronization of Events in Computer Networks 92
    5.2.1 Time-Based Multicast Protocols 92
    5.2.2 IC Model of Multicast Protocols 93
    5.2.3 Special Cases 94
    5.2.4 Specification of the Model 98
5.3 Virtual Clothing 100
    5.3.1 Problem Description 100
    5.3.2 An IC Scheme of Virtual Clothing 101

    5.3.3 Formalization 104
    5.3.4 Numerical Results 109
    5.3.5 Discussion 111

6 Applications 113

  6.1 Detecting Dangerous Situations in Near-Earth Space 113
  6.2 Complex Evaluation of Ecological and Medical Indicators 117
    6.2.1 Ecological Atlas of the City of Saint Petersburg 118
    6.2.2 The Ecological Atlas of the City of Kaliningrad 120
    6.2.3 Quality of Environment and Morbidity of Children for the City of Tula 122
    6.2.4 Similarity in Dynamics of Infection Morbidities in Russia 123
    6.2.5 Conclusions 126
  6.3 Surveillance of the Plague in Central Asia 127
    6.3.1 Plague Features 127
    6.3.2 AIS for Surveillance of the Plague 130
    6.3.3 Supervised Learning 132
    6.3.4 Unsupervised Learning 134
    6.3.5 Comparison with Traditional Statistics 136
    6.3.6 Conclusions 137
  6.4 Intelligent Security Systems 137
    6.4.1 Supervised Learning 141
    6.4.2 Unsupervised Learning 143

7 Immunocomputing Systems: Architecture and Implementation 145

  7.1 Toward an Immunocomputer 145
    7.1.1 Immunochip Architecture 145
    7.1.2 Implementation of Mathematical Operations 148
    7.1.3 Potential Applications for Information Security 154
  7.2 An Immunochip Emulator 165
    7.2.1 Version 1.0: Excitable Microarrays 165
    7.2.2 Version 1.1: 3D Excitable Microarrays 166
    7.2.3 Version 2.0: Self-Assembly of FP 167
    7.2.4 Version 3.0: 2D FIN 168
    7.2.5 Version 3.1: 3D FIN 171
    7.2.6 Version 3.2: Immune Response in 2D FIN 171
    7.2.7 Version 4.0: Orbital Hodograph 173
  7.3 The Biochip 173

7.3.1 The Biochip Matrix 175
7.3.2 The Molecular Adapter 176
7.3.3 The Biochip Controller 177
7.3.4 Biochip Applications... 180
7.3.5 From Biochip to Biocomputer... 181

Conclusion 183

Index 191

1
Introduction

1.1 The Mathematical Basis, Trends and Limitations of Modern Artificial Intelligence

Artificial intelligence (AI) first appeared as a scientific field within the overall framework of cybernetics and has, perhaps, suffered from the excessive optimism of its early stages. It seemed then that impressive results in the simulation of brain function would be reached through an increase in computer power. It was expected that the main hypothesis of AI, that "thinking is manipulation with symbols by prescript algorithms," would be confirmed (Kryukov, 1988) in the near future. This situation has been exacerbated by the usual practice of authors and publishers, who customarily publish successes but not failures, a practice that cannot give a realistic impression of the course of research.

Out of this background the so-called Lighthill's report, prepared by the famous mathematician J. Lighthill on behalf of the British Scientific Council, emerged in 1973. Not surprisingly, it received a sharply negative response from leading AI specialists, as well as criticism from within cybernetics (George, 1984). This report surveyed the field of AI, and separated it into two main directions. Lighthill treated the first direction as exclusively neurobiological, and called the second one "automation." The report acknowledged achievements on both fronts, but emphasized that its level of success had not lived up to the optimistic expectations of its early years. Although Lighthill failed to recognize the bridges that existed between the two areas of research, the main conclusions of his report are still justified.

Modern AI is clearly separated into the same two directions, in spite of various definitions. However, from our point of view, both directions are based on the modeling of natural systems, the most important of which are recognized as neural networks and languages. From this viewpoint these directions can be defined more formally as (1) logical–linguistic, or symbolic, and (2) biological.

The first direction is "classical" AI, in which the mathematical basis can be reduced to formal (mathematical) logic or to formal (generic) grammars.

2 Introduction

The second direction continues a long history of attempts to establish the basic principles of biological computing and to embody them in technical systems. Its most advanced fields are:

- artificial neural networks (ANN);
- cellular automata (CA);
- genetic algorithms (GA).

It is worth noting that the properties of AI systems are essentially defined by mathematical models of their basic elements and by the rules of interaction among the elements.

The basic element of ANN is an *artificial (or formal) neuron*, the output of which is usually computed as a function (threshold or sigmoid) of the sum of weighted inputs. It supposes that the weights of connections with other neurons, adjusted by ANN learning, simulate functions of the brain's memory (Wasserman, 1990). The notion of an artificial neuron was introduced as a result of studying the electrical activity of neural cells in an attempt to formulate their mathematical models. The results were based on the assumption that the electrical activity of the brain adequately represents the main biological mechanism of thinking.

ANN models are now widespread. They have a rigorous mathematical basis as well as hardware implementation in special *neurochips* and *neurocomputers*. Several kinds of ANN exist, and their learning algorithms (e.g., Perceptron, Boltzman machine, back propagation networks, Hopfield networks, Kohonen networks, etc.) are well known. However, these AI systems also have a number of serious drawbacks, including stable spurious patterns, nonlocalized errors, problems in the synchronization of events, representation of time, and representation of linguistic symbols.

Currently, ANN has moved away from its initial task of modeling its biological prototype to the realization of special physical and mathematical methods. Moreover, it has recently been concluded that the typical ANN tasks could be solved more efficiently by specialized mathematical methods (Rao Vemuri, 1992; Kanal, 1993).

The basic element of CA is the *cell*, which has a fixed set of states and a fixed position within the node of a spatial grid. A new state of any cell is computed as a function of the states of all its closest neighbors. It is also assumed that all cells change their states simultaneously in discrete time. This approach can be treated as a simple mathematical abstraction of the principles of interaction between biological cells. It has been shown (Toffoli and Margolus, 1987) that CA leads to the definition of a computing medium whose computational power is equivalent to the Turing machine. A serious problem of CA, though, is the lack of an adequate representation of phenomena by modeling (Adamatzky, 1994).

GA can be considered a result of the mathematical abstraction of the principles of natural selection on the molecular-genetic level (Forrest et al., 1993; Banzhaf and Reeves, 1999). The basic element of GA is the *bit string*, which represents a possible routine for solving some task. The obvious biological prototype of this element is DNA, which stores genetic code. GA uses a set or *population* of bit strings to solve tasks. The steps of GA consist in computing a *fitness function* for every string, deleting strings with the lowest value of this function, and simulating the genetic operations of *mutation* and *crossover* on the rest of the strings. GA has the same deficiencies as ANN and CA.

During the last two or three years, an active interest in the principles of information processing by the biological immune system, prompted by the fact that the immune system of vertebrates is much better understood than the human brain, has led to the development of a new field of computer science called artificial immune systems (AIS) (Dasgupta, 1999). The immune system possesses all the main properties of an AI system: ability to learn, ability to recognize, and ability to make decisions in its field of competence. At this point, AIS has not yet developed a proper concept or a mathematical definition of its basic elements.

From the above, the following two important conclusions can be derived:

First, we can see the importance of proper biological prototypes for AI, even on the rather primitive and abstract level of their current understanding and mathematical modeling. Moreover, such prototypes are able to help us develop new approaches to computing and even new types of computers.

Natural prototype	Biological level	AI model
Natural language	Left hemisphere of brain	Formal logic Formal linguistic
Neural networks	Cells	Neural computing
Biological cells	Cells	Cellular automata
Molecules of proteins	Molecular	?
Genetic code	Molecular	Genetic Algorithms

Table 1.1 Correspondence of AI models to their natural prototypes.

Second, if we consider the relationship between AI technologies and their biological prototypes, we can make an interesting observation. The rows in Table 1.1 list existing AI technologies and their biological prototypes in chronological order, from natural language to neural networks and genetic algorithms. However, another important biological prototype, molecules of proteins, which has been heretofore out of the scope of AI and computer science, should be placed between neural networks and genetic algorithms.

1.2 Key Mechanisms of Biomolecular Computing

It is a paradox that although the properties of single neurons have been more thoroughly studied than those of any other cell, our brain still remains the most mysterious organ. Our understanding of the brain is so rudimentary that we do not yet know whether the computer/brain analogy makes any sense at all. However, what we do know from modern molecular biology allows us to detect the surprising elegance and efficiency of all processes in living cells, as well as a remarkable consistency in the principles of their operation. There is no reason for us to expect that the brain will be an exception to these rules. Since all known attempts to understand intellect on the superficial level of neural networks have failed, it makes sense to concentrate our attention on the more profound functional levels.

1.2.1 Proteins from the Computing Viewpoint

From the computing viewpoint, all existing biological systems have a uniform information basis. It consists of the universal genetic code, and an "alphabet" of *amino acids*, where "words" are molecules of proteins, which are both the most complex of the known molecules and the most universal in their properties and functions. In computer terms, it could be said that the genetic code is analogous to software (instructions or programming that the cell receives from its parent cell), while the proteins could be viewed as hardware (the biophysical mechanisms that execute the program).

Proteins, being the neuro-mediators and the receptors of neurons, control the electrical activity of the brain. Proteins are also the main components of the immune system, which is composed of free proteins (*antibodies*, *messengers*, etc.) and proteins serving as *receptors* of *lymphocytes* (B-cells and T-cells). Therefore, proteins are expected to play key roles in both immunal and intellectual processes.

Although genes and proteins are exceptionally complex, some of their features can be explained by rather simple, generic mechanisms. These biophysical mechanisms, though, are not easy to uncover. A striking example is the discovery in 1953 of the double helix structure of the chain molecules that store the genetic code. This spatial structure is formed by the so-called *weak interactions* between very precisely determined molecular shapes situated in the same plane. This is one of the most significant examples of the geometrical correspondence among biomolecules. In the case of proteins, this level of understanding has yet to be reached. Nevertheless, the following principles are evident (Cantor and Schimmel, 1980a; B. Alberts et al., 1986; Bochinsky, 1987):

1. The spatial configuration of a protein is determined by the linear sequence (word) of its amino acids (letters);
2. This spatial configuration determines the function of any protein.

The first correspondence between the code and the stable configuration of a protein, its so-called *native form*, is ensured by the mechanisms of *self-assembly* or *folding*. The second correspondence between the spatial configuration and the function of a protein is implemented by the mechanisms of *molecular recognition*. Just as for the double helix, these mechanisms are essentially based on the weak interactions among different parts of the protein molecule and among different molecules of proteins.

Since modern mathematics possesses advanced means of description for both the spatial structures and for the sequences of symbols, such mechanisms of protein behavior seem especially promising for the development of new kinds of formal systems. Moreover, such systems seem to be capable of uniting in a natural way the two main directions of modern AI: biological and symbolic. Therefore, we will consider these mechanisms in more detail.

1.2.2 The Mechanisms of Protein Behavior

The main biophysical characteristic of weak interactions is *free energy* (Cantor and Schimmel, 1980b). The lower the energy, the stronger the interaction, and vice versa. Specifically, the negative energy corresponds to the attraction between molecules, while the positive energy corresponds to their repulsion. Thus, free energy is crucial in the definition of the main mechanisms of protein behavior, such as folding and recognizing, that are consistent with the reduction of the amount of free energy.

Self-Assembly

Self-assembly (or folding) is understood as the ability of the molecular chain of a protein to assume a single configuration in space, in spite of an astronomically large number of possible variants, through the interactions among its links. This ability of proteins to fold in unique ways gives them a remarkably precise way to adjust their structure to their function. Proteins are built so precisely that even a change of several atoms in a single link in the chain can violate the structure and lead to disastrous consequences.

The mechanisms of folding are closely connected to the difficult problem of the "extraction of order out of chaos," i.e., ensuring the unique properties of a protein molecule. This uniqueness is encoded in the alternation of flexible and hard links along the protein's chain, so-called *statistical balls* and *secondary structures*. The presence of these features ensures the folding of a protein's molecule, makes this folding quick and faultless, and provides the protein with necessary functional flexibility.

Recently, similar mechanisms of self-assembly have also been detected at the level of neural networks (Batuev and Babmindra, 1993). The formation of similar "topological maps" of neurons is being modeled by Kohonen networks (Haykin, 1999).

Molecular Recognition

Under molecular recognition (or just "recognition") we will be considering interactions with high specific selectivity between certain sectors of biomolecules. Such recognition forms the basis of several fundamental phenomena and is viewed as one of the main problems of biochemistry. The concept of biological specificity emerges as a result of molecular recognition. A similar phenomenon, crystallization, can be observed in inorganic nature. A growing crystal is able to "extract" certain molecules from a heterogeneous solution and "reject" all the others.

According to the "key and lock" hypothesis, such phenomena are determined by a mutual adjusting of molecules, like fragments in a mosaic. Such adjustments are based on weak interactions, which also determine the configuration of most biomolecules. In addition, molecular recognition depends more on sizes and shapes of interacting surfaces than on their chemical properties. However, the most recent experimental data indicates that mechanisms of molecular recognition are more complicated and cannot be explained merely by the shapes of the surfaces (Coutinho, 1994).

As a result of recognition, a protein can *bind* with another protein or molecule. Apparently, recognition/binding can be seen as the main information processing function of proteins. The protein is able to select or "recognize" the appropriate "pattern" as well as reject all the inappropriate ones. Moreover, such recognition can be directly quantified by the binding energy. As a result of binding, a protein can change its spatial configuration (shape). This phenomenon is called the *allosteric effect* (Cantor and Schimmel, 1980b). Furthermore, due to the allosteric effect the protein can sometimes bind with a protein (antigen, antibody, receptor, etc.) or another molecule with which it could not bind before. Thus, new proteins can be involved in subsequent binding, forming so-called *molecular circuits* and *immune networks*. It is worth noting that molecular circuits have been proposed as a possible molecular basis of neuronal memory in the human brain (Agnati, 1998).

1.2.3 The Immune System as a Model for a New Kind of Computing

Proceeding from the idea that all biological organisms have a uniform information basis, it follows that the mechanisms of intellect did not arise out of nothing. Apparently, they emerged from some more ancient mechanisms, just as the main mechanisms of the immune system evolved out of the more ancient system of cell recognition called *cell adhesion* (Edelman, 1989). Current understanding holds that the only valid candidate for the role of intellect's precursor is the immune system of vertebrates.

Such an assumption is unlikely to be thought extravagant. Three kinds of biological memory are already known: (a) genetic memory, (b) immune memory, and (c) usual memory. In spite of their obvious differences, experimental confirmations exist that all these forms of memory not only have common features, but also a common molecular mechanism (Rose, 1995; Agnati, 1998).

Actually, the immune system possesses not only memory, but is also able to learn, to recognize, and to make decisions about how to treat any molecule (antigen), even if that molecule has never before existed on Earth. Besides, the immune system defines the great diversity of possible molecular shapes in the highly individual context of its own experience, and it has developed strategies to free the genome from the task of straight coding such diversity.

These properties of the immune system allow specialists to call it "the second brain of vertebrates" (Coutinho, 1994). Moreover, in the phenomenon of immunity we encounter the basic problems of biology and biophysics. But if intellect remains a mystery, immunity is now comparatively well studied.

Immunity serves in the struggle against pathological microorganisms, and also ensures the genetic stability of cells. Besides this, the main task of the immune system is the elimination of mutated (specifically, cancer) cells and alien molecules. In response to the appearance of mutated cells and alien molecules, the immune system produces special cells and proteins (*antibodies*) that recognize the "enemy" and destroy it.

During the last few decades, the network theory of immunity has been acknowledged (Coutinho, 1995). N. Jerne, who worked at the Pasteur Institute of Paris, proposed the general theory of *idiotypic networks*, also called *immune networks* (Jerne, 1973, 1974). This theory is based on the concept that immune cells (*lymphocytes*) are not isolated, but communicate with each other. Among different species of lymphocytes such communication is accomplished through interaction among antibodies. Accordingly, the identification of antigens is achieved not by a single recognizing set but rather by a system-level recognition of the sets connected by the antigen–antibody reaction as a network.

The existence of such immune networks is now beyond all doubt, because their fragments and interactions have been detected experimentally. However, explanation of all the features of immunity is still beyond the network theory.

The ability of the immune system to generate a very accurate model of any biomolecule, facilitating recognition and binding, as well as the generation of antibodies, is more than a biological concept. It is especially interesting that antibodies are able to simulate the regulatory proteins of an organism's other systems, including the nervous system. During the last ten years numerous facts have appeared about the tightly reciprocal integration between nervous and immune systems called neural-immune modulation (Hori al., 1995; Korneva, 1996; Ader et al., 2000). These findings reveal the existence of a previously unknown form of computing performed at the level of the immune system.

We believe that the immune system presents a very interesting model for a new kind of computing, one that is likely to be no less useful than the model of the brain for the following reasons:

- Molecular immunology has developed as a leading direction in modern science (Mekler and Idlis, 1993), and the immune system is far better studied;
- The properties of the immune system seem to be a straight consequence of the basic biomolecular mechanisms of information processing by proteins;
- Formal description of the immune system can give qualitatively new mathematical and computer models and may also be a key to the mystery of the origin and embodiment of intellect.

1.3 Artificial Immune Systems

Modern biological principles of the immune system have provided the basis for a new and rapidly growing area of research called artificial immune systems (AIS). It could be said that AIS was "officially" established in 1999 when the fist book in this area was published (Dasgupta, 1999).

AIS represents a new and rapidly growing field of computer science, and it is expected to give rise to powerful and robust information processing capabilities for solving complex problems.

Apparently, AIS was prompted by the immune network theory of Jerne. His theory received a great deal of attention from immunologists and computer scientists. From the mathematical viewpoint, Jerne initiated the development of a rigorous framework for modeling the immune system. His theory is described by differential equations, which simulate the dynamics of concentration of lymphocytes.

Therefore, the 1970s marked the beginning of the vigorous development of mathematical modeling in immunology. However, this development was intended to describe the dynamics of the concentrations of the cells and proteins of immune systems by means of differential equations. Qualitatively, all such models are similar, differing only in terms of the number and order of equations, in the values of their coefficients, in the consideration of such factors as delays, thresholds or stochastic effects, and so on (Romanovsky et al., 1984; Perelson, 1989).

S. Forrest, who developed a *negative-selection algorithm* for change detection based on the principles of self–nonself discrimination in the immune system, is responsible for the second milestone in AIS (Forrest et. al, 1994).

This approach can be summarized as follows:

1. Define *self* as a collection S of strings of length l over a finite alphabet, a collection that needs to be protected or monitored. For example, S may be a normal pattern (program, data file) of activity, which is segmented into equal-sized substrings.
2. Generate a set R of *detectors*, each of which fails to match any string in S. Instead of exact or perfect matching, the method uses a *partial matching rule*, in which two strings match if and only if they are identical in at least r contiguous positions, where r is a suitably chosen parameter.
3. Monitor S for changes by continually matching the detectors in R against S. If any detector ever matches, then a change is known to have occurred, because the detectors are designed to match any of the original strings in S.

Other computational models exist that emulate different aspects of the immune system (e.g., Dasgupta and Attoh-Okine, 1997; Farmer et.al., 1986; Forrest et.al., 1993). Among them are the ability to detect common patterns in a noisy environment, to discover and maintain coverage of diverse pattern classes, and the ability to learn effectively in the environment when not all antibodies are expressed and not all antigens are presented.

The AIS approach can also be compared with the other AI approaches. Dasgupta (1997) has compared ANN and AIS and has found a number of similarities at the level of system behavior. Farmer et al. (1986), and Bersini and Varela (1990) compared the immune system with learning classifier systems. Gilbert and Routen (1994) experimented with immune network models to create a content-addressable autoassociative memory, specifically for image recognition.

It is worth noting that AIS has already been applied to several specific problems, such as information security, fault detection in manufacturing, vaccine design, data mining, and robotics. For example, Forrest and her group at the University of New Mexico are working on a research project with the long-term goal of building an AIS for computers (Forrest et al., 1997). Their computer immune system has to protect a computer against unauthorized use of computer facilities, maintain the integrity of data files, and prevent the spread of computer viruses. Their research program is based on the negative-selection algorithm.

Dasgupta and Forrest (1996) experimented with several time series data sets (both real and simulated) to investigate the performance of the negative-selection algorithm in detecting anomalies in the data series. The objective of this work was to develop an efficient algorithm that can be used for detecting any changes in the steady-state characteristics of a system or a process. In this case, the notion of self was considered to be the normal behavior patterns of the monitored system. Any deviation that exceeded an allowable variation in the observed data was considered an anomaly in the behavior pattern. The results have shown that this approach can be used as a tool for automated monitoring of safety-critical operations.

Ishida (1993) studied the mutual recognition feature of the immune network model for fault diagnosis. In his implementation, fault tolerance was attained by mutual recognition of interconnected units in the studied plant. That is, system-level recognition was achieved by unit level recognition. The results are very promising and worth further investigation.

Ishiguro et al. (1994) applied the immune network model to on-line fault diagnosis of plant systems. This work attempts to develop an integrated fault diagnosis method, which can be used in industrial plants.

Hunt and Cooke (1996) investigated an AIS based on the theory of immune networks within the context of machine learning. Such a system combines the advantages of learning classifier systems with some of the advantages of neural networks, machine induction, and case-based retrieval. They have shown the potential of AIS to work on pattern recognition problems, namely, the recognition of promoters in DNA sequences.

V. Skormin et al. (1999) formulated the *biological approach to system information security* as a generic approach aimed at the investigation and utilization of biologically inspired solutions to common computer network security problems (the BASIS project funded by the US Air Force).

These results show that there are many potential application areas in which AIS appears to be very useful. AIS is a subject of great research interest because of its powerful information processing capabilities. In particular, it performs many complex computations in a completely parallel and distributed fashion. Like ANN, AIS can learn new information, recall previously learned information, and perform pattern recognition tasks in a highly decentralized fashion. Also, learning takes place by evolutionary processes similar to evolutionary computations (Segel and Cohen, 2001).

However, unlike ANN, the field of AIS has not yet resulted in

- a proper mathematical basis, and
- hardware implementation analogous to existing neurocomputers.

Nowadays, AIS is represented by a number of heterogeneous software tools implementing heuristic algorithms, evolutionary computations, etc., and exploring various concepts "borrowed" from GA, CA, and ANN. In addition, there are as yet no models of immune networks that reflect the basic mechanisms of information processing by proteins. The first attempt to provide a uniquely specialized mathematical basis for the field of AIS was made only very recently (Tarakanov and Dasgupta, 2000).

1.4 Immunocomputing as a New Kind of Computing

This book is an attempt to fill the aforementioned gaps by presenting immunocomputing (IC) as a novel approach to computing that reflects recent discoveries in immunology. IC explores the principles of information processing that proteins and immune networks utilize in order to solve specific complex problems while protected from viruses, noise, errors, and intrusions. We intend to demonstrate that IC leads to a new kind of computer, which we propose to call an *immunocomputer* by analogy to the widely known *neurocomputers* that implement the models of neurons and neural networks.

Three main innovations are expected to emerge from IC:

1. A proper mathematical framework;
2. A new kind of computing;
3. A new kind of hardware.

These innovations determine the structure of this book.

The principal difference between IC and other kinds of computing stems from the functions of their basic elements defined according to their biological prototypes and their mathematical models (Tarakanov, 1998). For example, if an artificial neuron is considered as a summation with a threshold connected with fixed neurons (Wasserman, 1990), then the basic principles of IC are quite different: The main issues are free folding to a stable state and free binding with other elements dependent on their reciprocal states (see Section 1.2).

In fact, there are no mathematical models that can even approach these requirements. Thus, in Chapter 2 we develop a new concept, the *formal protein* (FP), as a mathematical abstraction for the key biophysical mechanisms of a natural protein's behavior. The FP has the same importance for IC as the artificial (or formal) neuron has for neural computing.

In the context of interaction between FPs we develop a new concept, *formal immune networks* (FIN), and demonstrate rigorously in the following chapters that such networks are able, like AI systems, to learn, recognize, and solve problems.

The closest mathematical models to FIN could be those based on Jerne's theory of idiotypic networks. As we mentioned in Section 1.3, his theory can also be formulated on the basis of differential equations representing the dynamics of the concentration of a set of lymphocyte clones and the corresponding immunoglobulins. However, such an approach does not consider the specific mechanisms of interactions between proteins and cells and therefore has limited applications.

As a new kind of computing, IC gives rise to the following innovations:

- An approach to pattern recognition and data mining based on the principles of biomolecular recognition (Chapter 3);
- An approach to language representation and problem solving based on an analogy between words and biomolecules (Chapter 4);
- An approach to modeling natural and technical systems based on the principles of biomolecular interactions (Chapter 5).

These approaches are demonstrated in Chapter 6, which considers a number of IC applications to solving specific complex real-life problems.

Chapter 7 lays the foundation for hardware implementation of IC in a special kind of electronic circuitry, since the architectures of traditional PCs and neurocomputers are not convenient for fast and distributed IC. Thus we propose an architecture for an immunochip, analogous to modern biochips, that could be developed into biomolecular computers in the future.

2
The Mathematical Basis of Immunocomputing

2.1 The Formal Protein

The notion of the *formal protein* (FP) abstracts the biophysical principle of free energy dependence from the spatial conformation of a protein. The notion of the FP unites two main ideas:

1. The algebraic description of a protein's chain geometry by means of quaternions;
2. The definition of free energy as a function over elements of quaternions.

2.1.1 The Algebraic Description of the Shape of Natural Proteins

According to (Cantor and Schimmel, 1980a), the spatial conformation of a protein's skeleton can be represented geometrically as in Figure 2.1, where k is a number of repeated sections.

The bold letters in Figure 2.1 designate the following atoms: $\mathbf{C^a}$, alpha carbon; $\mathbf{C}$, carboxyl carbon; $\mathbf{N}$, nitrogen; $\mathbf{O}$, oxygen; $\mathbf{H}$, hydrogen. The spatial conformation of the protein is determined by the fixed *valence angles* θ, η, ξ of the valence bonds between the atoms $\mathbf{N}$, $\mathbf{C^a}$, $\mathbf{C}$, and by the *torsion angles* of rotation φ, ψ of the skeleton around the bonds $\mathbf{N} - \mathbf{C^a}$ and $\mathbf{C^a} - \mathbf{C}$.

As shown in Figure 2.1, the Cartesian system *XYZ* is used for the computation of the coordinates of the protein's atoms. The *X*–axis of this system represents the valence bond $\mathbf{N} - \mathbf{C^a}$; the *Y*–axis is situated in the plane $\mathbf{H} - \mathbf{N} - \mathbf{C^a}$ and is directed toward $\mathbf{H}$; and the *Z*–axis serves as the third system coordinate, directed to the right.

14 The Mathematical Basis of Immunocomputing

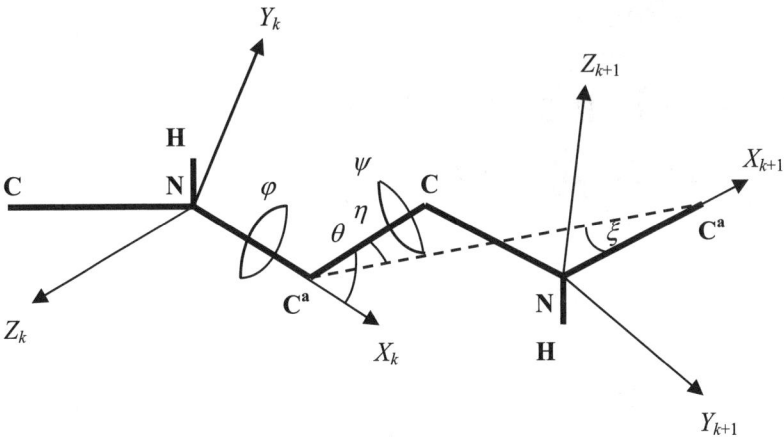

Figure 2.1 The spatial configuration of a protein's skeleton.

According to the proximity of the experimental values of the angles η and ξ (Cantor and Schimmel, 1980a), let us assume that $\eta=\xi$, and let us also introduce the following designations:

$$c_0 = \cos\frac{1}{2}\theta, \quad s_0 = \sin\frac{1}{2}\theta, \quad c_1 = \cos\frac{1}{2}\eta, \quad s_1 = \sin\frac{1}{2}\eta,$$

$$\sigma = \frac{1}{2}(\psi + \varphi), \quad \delta = \frac{1}{2}(\psi - \varphi).$$

Then the transition between the systems $(XYZ)_k$ and $(XYZ)_{k+1}$ for any k can be described algebraically by the following quaternion:

$$Q(\varphi, \psi) = q_0 H_0 + q_2 H_1 + q_3 H_2 + q_4 H_3, \tag{2.1}$$

$$\begin{aligned}
q_1 &= -c_0 \sin\sigma, \\
q_2 &= c_0 c_1 \cos\sigma + s_0 s_1 \cos\delta, \\
q_3 &= c_0 s_1 \cos\sigma - s_0 c_1 \cos\delta, \\
q_4 &= -s_0 \sin\delta,
\end{aligned} \tag{2.2}$$

where H_0, H_1, H_2, H_3 are Pauli matrices (Casanova, 1976):

$$H_0 = \begin{bmatrix} 1 & 0 & 0 & 0 \\ 0 & 1 & 0 & 0 \\ 0 & 0 & 1 & 0 \\ 0 & 0 & 0 & 1 \end{bmatrix}, \quad H_1 = \begin{bmatrix} 0 & 0 & 0 & -1 \\ 0 & 0 & 1 & 0 \\ 0 & -1 & 0 & 0 \\ 1 & 0 & 0 & 0 \end{bmatrix},$$

$$H_2 = \begin{bmatrix} 0 & 0 & 1 & 0 \\ 0 & 0 & 0 & 1 \\ -1 & 0 & 0 & 0 \\ 0 & -1 & 0 & 0 \end{bmatrix}, \quad H_3 = \begin{bmatrix} 0 & -1 & 0 & 0 \\ 1 & 0 & 0 & 0 \\ 0 & 0 & 0 & 1 \\ 0 & 0 & -1 & 0 \end{bmatrix}.$$

These matrices have the following properties:

$$H_1 H_2 = -H_3, \quad H_2 H_3 = -H_1, \quad H_3 H_1 = -H_2, \quad H_i^2 = -H_0, \quad H_i H_j = -H_j H_i,$$
$$i, j = 1, 2, 3, \quad i \neq j.$$

Let us also designate the quaternion $Q(\varphi, \psi)$ as

$$Q(\varphi, \psi) = (q_1, q_2, q_3, q_4).$$

The consecutive multiplication of such quaternions allows us to compute the coordinates of any atom of a protein. Thus, we obtain the algebraic description of the natural 3D shape of a protein.

2.1.2 The Mathematical Notion of a Formal Protein

Definition 2.1 A *formal protein* is an ordered 5-tuple

$$P = \langle n, U, Q, V, v \rangle,$$

which comprises the following components:

1. The number of links $n > 0$;
2. The set of angles $U = \{\varphi_k, \psi_k\}$, $k = 1, \ldots, n$, where $-\pi \leq \varphi_k \leq \pi, -\pi \leq \psi_k \leq \pi$;
3. The set of unit quaternions $Q = \{Q_0, Q_k\}$, where quaternions $Q_k = Q_k(\varphi_k, \psi_k)$ are defined by formulas (2.1) and (2.2), and a *resultant quaternion* of FP Q_0 is defined as their product:

$$Q_0 = Q_1 Q_2 \cdots Q_n;$$

16 The Mathematical Basis of Immunocomputing

4. The set of coefficients $V = \{v_{ij}\}$, $i = 1, 2, 3, 4$, $j \geq i$;
5. The function v (without index), is defined over the elements of the resultant quaternion Q_0 by the following quadratic form:

$$v = -\sum_{j \geq i} v_{ij} q_i q_j. \tag{2.3}$$

Let us consider the values of the torsion angles U as *states* of an FP, and the values of the coefficients V as *controls* of an FP.

Introduce the *Q-vector* of an FP as a column vector $[Q] = [q_1, q_2, q_3, q_4]^T$, where "$T$" is the transpose symbol. The coordinates of the Q-vector are the elements of the resultant quaternion Q_0. Consider the *energy matrix* $M(v)=[m_{ij}]$, $i,j = 1,2,3,4$, determined by the coefficients of the quadratic form (2.3) to be symmetric and defined as follows:

$$m_{ii}=v_{ii},\ m_{ij}=m_{ji}=\frac{1}{2}v_{ij},\ i \neq j.$$

Hence, the function (2.3) can be represented in vector–matrix form:

$$v = -[Q]^T M(v)[Q]. \tag{2.4}$$

Definition 2.2 The *free energy* of an FP is a quantity determined by the quadratic form (2.4).

Later we will also use *energy* as a synonym for the free energy of an FP.

Definition 2.3 The *self-assembly* of an FP is a process whereby the FP changes its state according to the following system of differential equations:

$$\begin{aligned}\ddot{\varphi}_k &= -\frac{\partial}{\partial \varphi_k} v - c\dot{\varphi}_k + f(h), \\ \ddot{\psi}_k &= -\frac{\partial}{\partial \psi_k} v - c\dot{\psi}_k + f(h),\end{aligned} \tag{2.5}$$

where $k=1,\ldots,n$.

Equations (2.5) describe the dynamics of state change for bodies with unit masses under the influence of potential v, resistance of environment, and disturbances. Here c is a coefficient of viscous friction, and $f(h)$ is a function that defines stochastic disturbances whose intensity is quantified by the parameter h. This parameter represents such factors as the temperature or pH of the environment, as well as any other external factors.

From the point of view of computing, we are interested not in the process of self-assembly itself, but in the formation of stable states of FP. Such states could correspond to the storing and retrieving of information, as well as to recognition among FPs. Evidently, from (2.5), with the absence of external disturbances, stable states of an FP and transitions between them are determined by the stationary values of the free energy, where its partitive derivation over any torsion angle is equal to zero. In addition, stable states are the stationary states, which correspond to the local minima of the free energy. Thus, stable states are determined by the extrema of the quadratic form (2.4).

The solution of this problem would be straightforward if extrema to be determined were dependent only on the elements of the resultant quaternion. Then the solution, consistent with the representation (2.4), could be determined by the spectral factorization of the energy matrix, formed by an FP's controls. However, solutions of (2.5) ought to be determined relative to the torsion angles, and the free energy of an FP has nonlinear dependence on these angles. Moreover, (2.2) determines nonlinear boundaries for the possible values of elements Q_0. Thus, special methods have been developed (in Tarakanov, 1999a) for the study of stationary states of an FP.

Therefore, from a computing viewpoint, an FP is described by four real numbers (resultant quaternion), which are computed as a product of elementary quaternions and are analogous to the amino-acid links of the native proteins. Every elementary quaternion, in turn, is computed as a function of two torsion angles of rotation. The free energy of an FP is computed as a sum of mutual products of resultant quaternion elements with given weight coefficients. From a control systems viewpoint, the torsion angles of an FP correspond to its inner parameters (states), while the weight coefficients of the quadratic form (2.3) function as the external parameters (control efforts). From both viewpoints, control efforts constitute the input of an FP, while its states are an output.

Apparently, the FP is the simplest possible mathematical model that demonstrates such important features of the native protein as self-assembly and noncommutative dependence of the native form upon the number and the order of links.

2.2 The Interaction Between Formal Proteins

2.2.1 Binding Energy

According to Section 1.2.2, the main condition for a protein to function is binding with another protein or another molecule. The main biophysical characteristic of binding between proteins is also free energy: The lower the energy, the stronger the binding, and vice versa. Thus, negative energy, lower than the energy of Brownian motion, corresponds to proper binding, while positive energy corresponds to repulsion between proteins.

Accordingly, let us consider the free energy of interactions between FPs as a *binding energy*, to distinguish it from the free energy of a protein's folding. As a mathematical abstraction of this fact, consider a natural spread of the quadratic form (2.4) on the interaction between FPs.

Definition 2.4 The *binding energy* between two FPs is a bilinear form:

$$w(P,Q) = -[P]^T W[Q], \qquad (2.6)$$

where $[P]$, $[Q]$ are Q-vectors of the first and the second FPs, respectively, W is a binding matrix, $W = \{w_{ij}\}$, where w_{ij} are given coefficients, and $i,j = 1,2,3,4$.

This definition indicates that $w(Q,P) = -[Q]^T W^T[P]$, and comparing bilinear forms, determined by the matrices W, W^T, provides the evidence that for any two configurations $w(P,Q) = w(Q,P)$. In other words, the binding energy of two FPs is invariant to permutations of the configuration vectors in (2.6).

Definition 2.5 *Binding* is an interaction between two FPs with binding energy $w \leq w_h$, where w_h is some *threshold of binding*. If this condition is satisfied, then FPs are considered bound FPs, and the following assumptions can be made:

1. Bound FPs form a complex unit;
2. The energy of the complex is equal to the sum of its free energies plus binding energy:

$$v(P,Q) = v(P) + v(Q) + w(P,Q); \qquad (2.7)$$

3. The self-assembly of each bound FP is determined by the energy of the complex (2.7).

2.2.2 The Allosteric Effect

It is evident from (2.7) that the stable states of bound FPs depend on the energy of their complex. Therefore, it can be shown that an FP is able to transfer to another stable state by binding with another FP.

Definition 2.6 The *formal allosteric effect* is the changing of the stable state of an FP as a result of binding.

Proposition 2.1 The allosteric effect exists.

Example 2.1 Consider an FP with $n=1$ (*monopeptide*), $\theta = \eta = 0$, $\varphi = -\pi$. Designate its quaternion as $Q_x(\psi)$. Then, according to (2.2),

$$Q_X(\psi) = \left(\cos\frac{\psi}{2}, \sin\frac{\psi}{2}, 0, 0\right), \quad -\pi \leq \psi \leq \pi, \quad \cos\frac{\psi}{2} > 0.$$

Let this FP have a single nonzero control v_{11}. Then the energy of the FP and its derivatives are as follows:

$$v = -v_{11}\cos^2\frac{\psi}{2}, \quad v'_\psi = \frac{1}{2}v_{11}\sin\psi, \quad v''_\psi = \frac{1}{2}v_{11}\cos\psi.$$

Hence, when $v_{11} > 0$ such an FP has a single stable state $\psi^* = 0$, and when $v_{11} < 0$, $\psi^* = \pi$.

Consider an interaction between two such FPs $Q_X(\psi_1)$, $Q_X(\psi_2)$. Let each of the FPs have a single nonzero control, which is designated v_1 for the first FP and v_2 for the second FP. Let the binding energy be determined by a single nonzero coefficient w_{11}. Then it can be shown (Tarakonov, 1999a) that the energy and its partial derivatives for such a complex are given by the following equations:

$$v = -v_1\cos^2\frac{\psi_1}{2} - v_2\cos^2\frac{\psi_2}{2} - w_{11}\cos\frac{\psi_1}{2}\cos\frac{\psi_2}{2},$$

$$\frac{\partial v}{\partial \psi_i} = \left(v_1\cos\frac{\psi_1}{2} + \frac{w_{11}}{2}\cos\frac{\psi_{3-i}}{2}\right)\sin\frac{\psi_i}{2}, \quad i = 1, 2,$$

$$\frac{\partial^2 v}{\partial \psi_i \partial \psi_i} = \frac{v_i}{2}\cos\psi_i + \frac{w_{11}}{4}\cos\frac{\psi_1}{2}\cos\frac{\psi_2}{2},$$

$$\frac{\partial^2 v}{\partial \psi_i \partial \psi_{3-i}} = -\frac{w_{11}}{4}\sin\frac{\psi_1}{2}\sin\frac{\psi_2}{2}.$$

Let $v_1 = 1$, $v_2 = -1$.

Consider first the case $w_{11} = 0$, when the FPs do not interact. In this case each FP has a single stable state $\psi_1^* = 0$, $\psi_2^* = \pi$.

Let $w_{11} = 1$. Then necessary and sufficient conditions of the extrema give the following stable states of the complex: $\psi_1^* = 0$, $\psi_2^* = \pm\frac{2\pi}{3}$. Hence, after binding, the stable state of the first FP does not change, while the second FP gets two new stable states.

This example proves Proposition 2.1.

2.2.3 Networks of Binding

The allosteric effect can give an FP the ability to bind to itself another FP that initially could not be bound. The resultant FPs can be involved in the process of subsequent binding.

Definition 2.7 The *network of binding* is a sequence of bindings between FPs that includes allosteric effects.

Proposition 2.2 Networks of binding exist.

Example 2.2 Consider an FP with $n=1$, $\theta=\eta=0$, $\psi=\varphi-\pi$. Designate its quaternion as $Q_d(\varphi)$. Then, according to (2.2),

$$Q_d(\varphi) = (\cos \varphi, \sin \varphi, 0, 0).$$

Consider a system of three such FPs: $\{P_1, P_2, P_3\}$. Let them exist in their stable states. Correspondingly (see the bold lines in Figure 2.2a),

$$\varphi_i^* = \frac{2\pi}{3}(i-1), \quad i=1,2,3.$$

Assume that by binding, any of these FPs can change its stable state to a new one:

$$\varphi_i^{**} = \varphi_i^* + \pi.$$

Define the threshold of binding $w_h = 0$ and the energy of interaction between P_i and P_j as:

$$w(P_i, P_j) = -\cos\left(\varphi_i^* - \varphi_j^*\right).$$

Then no one pair of given FPs is able to bind.

Let a fourth FP, denoted by A (antigen), be added to the system. Let this FP exist in its stable state $\varphi_4^* = \frac{\pi}{3}$ (dotted lines in Figure 2.2), so that it can bind only P_1. Consider the following events:

- A binds P_1 (Figure 2.2a), and P_1 changes its state so that P_1 can bind P_2 or P_3;
- P_1 binds P_2 (Figure 2.2b), and P_2 changes its state so that P_2 can bind P_3;
- P_2 binds P_3 (Figure 2.2c), and P_3 changes its state so that P_3 can bind A (Figure 2.2d), etc.

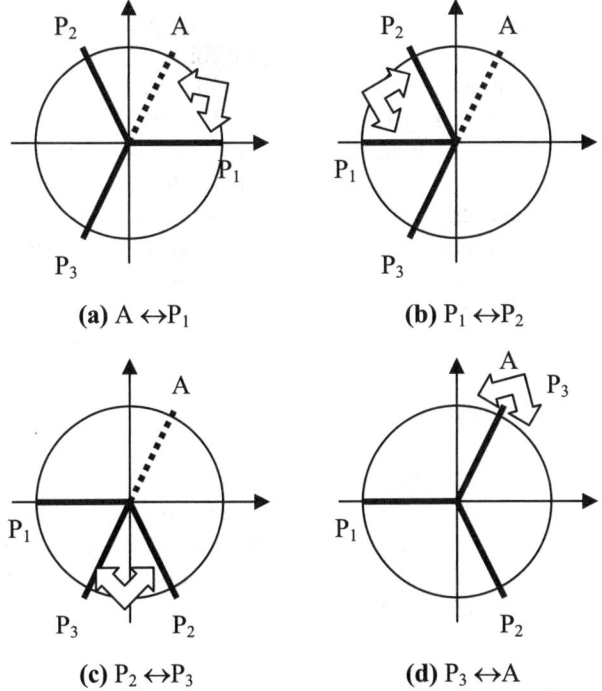

Figure 2.2 Diagram of bindings between FPs.

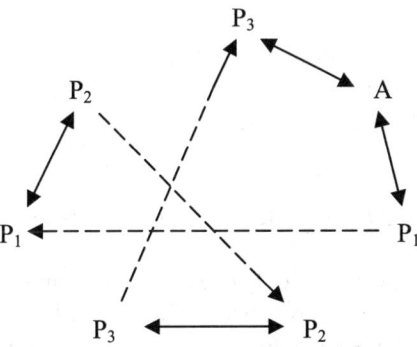

Figure 2.3 Schematic representation of the network of binding.

The described network of bindings between FPs {A, P₁, P₂, P₃} is shown schematically in Figure 2.3, where dotted lines designate transitions between stable states of an FP. Obviously, other ways of binding also exist between the given FPs. For example, the initial activation of system {P₁, P₂, P₃} by the antigen A could be realized by binding A↔P₂ (Figure 2.2a). Figure 2.2b,c also shows that several variant bindings exist, but the condition of the system, corresponding to Figure 2.2d, shows only one possible way of binding.

Thus, it can be said that the antigen launches a network of binding between FPs: {A, P₁, P₂, P₃}. Moreover, it can also be said that the network trains P_3 to recognize the antigen. From the computing viewpoint this means that a network of binding is able to "learn."

2.3 Formal Immune Networks

For mathematical modeling of immune networks, we need to supply the network of binding between FPs with models of immune cells, including their proliferation and death. For this purpose we introduce notions of the *formal B-cell* and the *formal T-cell* and define FIN as networks of binding that include bindings of *B-cells* and/or *T-cells*.

2.3.1 The Formal B-Cell

Definition 2.8 The *Formal B-cell* is an ordered quadruplet

$$B\text{-}cell = \langle P, Ip, Is, Im \rangle,$$

where

- *P* is a *cell receptor* that is also an FP;
- *Ip* is a *receptor state indicator*;
- *Is* is a *cell state indicator*;
- *Im* is a *mutation indicator*.

The behavior of a *B-cell* is defined by the following rules:

1. A *B-cell* can exist in only one of the following states: $Is = \{0, 1, 2\}$:
   $Is = 0$ corresponds to *death* when a *B-cell* is destroyed;
   $Is = 1$ corresponds to *recognition* when the receptor *P* of the *B-cell* can be bound with any other FP;
   $Is = 2$ corresponds to cell *proliferation* when a *B-cell* is divided into two copies such that for both, the cell states are $Is = 1$ and the receptor states are determined by the *Im*;

Transition from the state *Is*=1 to the state *Is*=2 occurs only as a result of the binding of the receptor *P* with any other FP.

2. The mutation indicator *Im* = {0, 1} of the parent *B-cell* determines the states of the receptors of its copies as follows:

Im=0: receptors inherit their types from the parent *B-cell* (*without mutation*);

Im=1: receptors change their types (*with mutation*).

In other words, the main property of a formal *B-cell* is its proliferation (or death) as a result of free binding (or nonbinding) with any FP.

2.3.2 The Formal T-Cell

Definition 2.9 The *Formal T-cell* is an ordered set

$$T\text{-}cell = \langle K, P_0, P_1, \ldots, P_k \rangle,$$

where

- *K* is a type of *T-cell*;
- $P_1, \ldots, P_k$ are receptors of a *T-cell*, viewed as FPs;
- P_0 is an FP, which is secreted by a *T-cell* if all its receptors are bound.

In other words, the main function of a formal *T-cell* is to secrete an FP of a specific type when each of its receptors becomes bound by an FP.

This behavior of a *T-cell* can also be described as a rule of an attributive formal grammar (Tarakanov, 1999a):

$$P_0 \rightarrow \langle K \rangle P_1 \ldots P_k.$$

This representation allows us to develop a specific kind of FIN with *T-cells* as a mathematical model for language representation and knowledge-based reasoning.

2.3.3 The Mathematical Notion of FIN

Definition 2.10 A *Formal Immune Network* (FIN) is a network of binding between FPs, including receptors of *B-cells* and/or receptors of *T-cells* and/or *FFPs* (Free FPs), which are not receptors of any cell:

$$\text{FIN} = \langle B\text{-}cells \cup T\text{-}cells \cup FFPs \rangle.$$

According to this definition, an FIN should include at least *B-cells*, while a network of binding can include only *FFPs*.

Unlike cellular automata or ANNs with fixed elements and connections, cells of FINs are allowed to change their places and to bind freely with each other by means of receptors.

Integer-Valued FIN (IFIN)

Definition 2.11 An *IFIN* (n, n_h) satisfies the following rules:

1. $Ip = \{0, 1,..., n-1\}$ for every *B-cell*; accordingly, designate the types of receptors and *FFPs* as $P(0), P(1), ... , P(n-1)$, the types of *B-cells* as $B(0), B(1), ... , B(n-1)$, and the types of *T-cells* as $T(0), T(1), ..., T(n-1)$;
2. Binding energy is defined by the formula

$$w(P(i), P(j)) = \min \{ (i-j)\mod(n), (j-i)\mod(n) \};$$

3. Binding occurs if and only if $w \leq n_h$, where n_h is an integer threshold of binding;
4. A *B-cell* proliferates if and only if it binds with any neighbor; otherwise, a *B-cell* dies;
5. All *B-cells* proliferate or die simultaneously in discrete time steps.

We have introduced and studied theoretically two kinds of IFIN: the so-called BB-networks and AB-networks (Tarakanov, 1999a).

The BB-Network

The BB-network represents a case where several types of *B-cells* are generated and self-maintained by interactions among the *B-cells* themselves, in the absence of any external *FFPs*.

Definition 2.12 A one-dimensional BB-network $1DBB(n, n_h)$ is an IFIN that contains only *B-cells* with the following rules (in addition to Definition 2.11):

1. *B-cells* form a 1D sequence (*population*) without gaps, with the first cell B_1 and the last cell B_m:

$$B_1,..., B_{k-1}, B_k, B_{k+1},...,B_m ;$$

2. Every inner *B-cell* B_k has an *eastern neighbor* B_{k+1} and a *western neighbor* B_{k-1}; accordingly, the first cell has only an eastern neighbor, and the last cell has only a western neighbor;

3. If cell B_k proliferates, then its copies B_{k1} and B_{k2} distribute themselves between the neighbors of the parent cell:

$$B_1,..., B_{k-1}, B_k, B_{k+1},...,B_m \rightarrow B_1,..., B_{k-1}, B_{k1}, B_{k2}, B_{k+1},...,B_m\ ;$$

4. If cell B_k dies, then an empty place $\varnothing$ remains:

$$B_1,..., B_{k-1}, B_k, B_{k+1},...,B_m \rightarrow B_1,..., B_{k-1}, \varnothing, B_{k+1},...,B_m\ ;$$

5. After every step, the population shifts to fill all the empty places.

The AB-Network

Definition 2.13 A one-dimensional AB-network $1DAB(n, n_h)$ is an IFIN that contains *B-cells* and *FFPs* (*antigens*) of n types with the following rules:

1. The population of *antigens* is placed above the population of *B-cells* so that every *B-cell* has no more than one matching *antigen* above:

$$A_1,..., A_k,$$
$$B_1,..., B_k,...,B_m;$$

2. A *B-cell* dies if it has no matching *antigen* or if $w > n_h$;
3. If $w = n_h$, then a *B-cell* proliferates by two precise copies of itself (without mutations);
4. If $w < n_h$, then a *B-cell* proliferates by two copies of its nearest types (with mutations);
5. The population of *antigens* remains invariable.

The Two-Dimensional FIN (2DFIN)

A two-dimensional BB-network represents a natural spread distribution of a 1D network. However, the topology of 2D and 3D spaces differs from the simple topology of a 1D space. Thus specific rules are needed for arranging proliferated cells and filling the gaps from the dead cells.

Definition 2.14 A two-dimensional BB-network $2DBB(n, n_h)$ is an IFIN that contains only *B-cells* with the following rules:

1. *B-cells* form a 2D population:

$$B_{1,1},..., B_{1,j},...,$$
$$...,$$

26 The Mathematical Basis of Immunocomputing

$$B_{i,1}, \ldots, B_{i,j}, \ldots,$$
$$\ldots;$$

2. Every inner B-cell $B_{i,j}$ has four neighbors: *east* $B_{i,j+1}$, *west* $B_{i,j-1}$, *north* $B_{i+1,j}$, and *south* $B_{i-1,j}$; accordingly, bordering B-cells have a reduced number of neighbors;
3. Inherited copies $B(k_1)$, $B(k_2)$ of a cell $B_{i,j}(k)$ arrange themselves between the nearest neighbors by the following rule, where k is the type of the parent B-cell and k_1, k_2 are mutated types:

$B_{i-1,j}$		$B_{i-1,j+1}$
$B(k_1)$	$B(k_2)$	
$B(k_2)$	$B(k_1)$	
$B_{i+1,j}$		$B_{i+1,j+1}$

4. After every step, the population shifts to fill all empty rows and columns.

The Three-Dimensional FIN (3DFIN)

Definition 2.15 A three-dimensional BB-network $3DBB(n, n_h)$ is an IFIN that contains only B-cells with the following rules:

1. B-cells form a 3D population, where r is a number of the horizontal layer:

$$B_{1,1,r}, \ldots, B_{1,j,r}, \ldots,$$
$$\ldots$$
$$B_{i,1,r}, \ldots, B_{i,j,r}, \ldots,$$
$$\ldots;$$

2. Every inner B-cell, $B_{i,j,r}$ has six neighbors: *east* $B_{i,j+1,r}$, *west* $B_{i,j-1,r}$, *north* $B_{i+1,j,r}$, *south* $B_{i-1,j,r}$, *up* $B_{i,j,r+1}$, and *down* $B_{i,j,r-1}$; accordingly, bordering B-cells have a reduced number of neighbors;
3. Inherited copies $B(k_1)$, $B(k_2)$ of a cell $B_{i,j,r}(k)$ arrange themselves between the nearest neighbors by the following rule, where k is the type of the parent B-cell and k_1, k_2 are mutated types:

$B_{i-1,j,r}$		$B_{i-1,j+1,r}$
$B(k_1)$	$B(k_2)$	
$B(k_2)$	$B(k_1)$	
$B_{i+1,j,r}$		$B_{i+1,j+1,r}$

$B_{i-1,j,r+1}$			$B_{i+1,j+1,r+1}$
$B(k_2)$	$B(k_1)$		
$B(k_1)$	$B(k_2)$		
$B_{i+1,j,r+1}$			$B_{i+1,j+1,r+1}$

4. After every step, the population shifts to fill all empty layers by i, j, and r.

2.3.4 The Properties of FINs

The following results have been proved for 1DFINs in (Tarakanov, 1999a).

Theorem 2.1 If all *antigens* in an $AB(n, n_h)$ network are of the same type, and at least one *B-cell* binds an *antigen*, then after a finite number of steps, for every *antigen* there will correspond a matching *B-cell*.

Theorem 2.2 For any initial population of any $BB(n, n_h)$ network, only one of the following three modes is possible:

1. Death of all B-cells (*formal immunodeficiency*);
2. Unlimited proliferation of B-cells (*formal allergy*);
3. Cyclic reproduction of the initial population (*formal immune memory*).

Theorem 2.3 For any n there exists a threshold n_h such that at least one cyclic mode is possible in the $BB(n, n_h)$ network.

Theorem 2.1 shows how *antigens* can control the reproduction and death of *B-cells* even within the simplest variant of FINs. Apparently, this result represents a simplest model of *formal immune response*, which implies the intention of *B-cells* to accept a type of maximal affinity to the *antigen*.

Theorems 2.2 and 2.3 represent a mathematical model of self-maintaining immune memory, where several types of *B-cells* are generated and stored by interactions between *B-cells* themselves, in the absence of any external *antigen*. Computer modeling shows that in any $BB(n, n_h)$ network with a rather large n, there exists a large number of cyclic modes with several periods and different population sizes.

In general, the obtained results show that even the simplest variants of FINs demonstrate such important effects as the following:

- immune response under the control of an antigen;
- immune memory and generation of a new immune repertoire in the absence of an outer antigen.

However, if the notion of 1DFINs still yields to a pure mathematical description, then 2DFINs and especially 3DFINs are already much fuzzier. Research into their properties is practically impossible without computer modeling. At the same time, these FINs seem to be the closest mathematical models, respectively, to the modern 2D biochip microarrays (Ekins and Chu, 1999; MacBeath and Schreiber, 2000) and to the natural 3D immune system (Coutinho, 1994).

2.4 Related Topics

2.4.1 Quaternion Algebra

Using Pauli matrices H_0, H_1, H_2, H_3 (see Section 2.1.1) every quaternion

$$Q = q_1 H_0 + q_2 H_1 + q_3 H_2 + q_4 H_3$$

can be represented as a matrix, where q_1, q_2, q_3, q_4 are real values:

$$Q = \begin{bmatrix} q_1 & -q_4 & q_3 & -q_2 \\ q_4 & q_1 & q_2 & q_3 \\ -q_3 & -q_2 & q_1 & q_4 \\ q_2 & -q_3 & -q_4 & q_1 \end{bmatrix}.$$

However, from a computing viewpoint it is convenient to introduce and operate on quaternions defined as quadruplets of real numbers:

$$Q = (q_1, q_2, q_3, q_4).$$

The basis of quaternion algebra is represented by Pauli matrices, which determine the following properties of quaternions:

Two quaternions Q and P are equal if their elements are equal

$$q_i = p_i, \; i = 1,2,3,4.$$

The sum of the quaternions Q and P is a quaternion, whose elements are represented as follows:

$$q_i + p_i : Q + P = (q_1 + p_1)H_0 + (q_2 + p_2)H_1 + (q_3 + p_3)H_2 + (q_4 + p_4)H_3.$$

The multiplication of a quaternion Q by a scalar a is multiplication by this number for each of its elements:

$$aQ = aq_1H_0 + aq_2H_1 + aq_3H_2 + aq_4H_3.$$

In particular, the quaternion

$$-Q = -q_1H_0 - q_2H_1 - q_3H_2 - q_4H_3$$

is the additive inverse of Q, and the zero quaternion is as follows: (0, 0, 0, 0).

The addition of quaternions and multiplication of them by a scalar submits to the usual rules of algebra:

$$Q + P = P + Q; \quad (Q + P) + S = Q + (P + S);$$
$$aQ = Qa, \quad (ab)Q = Q(ba);$$
$$(a + b)Q = aQ + bQ, \quad a(Q + P) = aQ + aP.$$

It is possible to consider the basic elements, Pauli matrices H_0, H_1, H_2, H_3, as basis vectors of four-dimensional space, designated as **H**. Then any quaternion can be presented in this space by a point or radius vector. The addition of vectors and multiplication of them by a scalar in the space **H** occurs in the same way as in a usual vector space. A useful feature of the space **H** is that it is closed concerning the multiplication and division operations.

To define a product of quaternions, it is necessary to set rules of multiplication of Pauli matrices H_0, H_1, H_2, H_3. These rules are as follows:

$$H_0H_1 = H_1H_0,$$
$$H_0H_2 = H_2H_0,$$
$$H_0H_3 = H_3H_0,$$
$$H_0H_0 = H_0,$$
$$H_1H_1 = -H_0,$$
$$H_2H_2 = -H_0,$$
$$H_3H_3 = -H_0,$$
$$H_1H_2 = -H_2H_1 = H_3,$$
$$H_3H_1 = -H_1H_3 = H_2.$$

With this rule of multiplication, the product of two quaternions is again a quaternion.

A quaternion may be presented as a sum of *scalar* and *vector* parts, which are labeled as scal Q and vect Q accordingly:

$$Q = \text{scal } Q + \text{vect } Q = q_1 + q.$$

The product of two quaternions Q and P is given by

$$QP = (q_1p_1 - q_2p_2 - q_3p_3 - q_4p_4)H_0$$
$$+ q_1(p_2H_1 + p_3H_2 + p_4H_3) + p_1(q_2H_1 + q_3H_2 + q_4H_3)$$
$$+ (q_3p_4 - q_4p_3)H_1 + (q_2p_4 - q_4p_2)H_2 + (q_2p_3 - q_3p_2)H_3.$$

The products of three Pauli matrices are equal to

$$H_1H_2H_3 = H_3H_3 = -H_0, \quad H_1H_2H_3 = H_1H_1 = -H_0,$$
$$H_1H_3H_2 = -H_2H_2 = H_0, \quad H_1H_3H_2 = -H_1H_1 = H_0,$$
$$H_1H_2H_1 = H_3H_1 = H_2, \quad H_1H_2H_1 = -H_1H_3 = H_2,$$
$$H_1H_1H_2 = -H_2, \quad H_1H_1H_2 = H_1H_3 = -H_2,$$
$$H_1H_3H_1 = -H_2H_1 = H_3, \quad H_1H_3H_1 = H_1H_2 = H_3,$$
$$H_1H_1H_3 = -H_3, \quad H_1H_1H_3 = -H_1H_2 = -H_3.$$

The multiplication of quaternions has associative and distributive properties in relation to addition:

$$(QP)S = Q(PS);$$
$$Q(P + S) = QP + QS.$$

The multiplication of quaternions is generally noncommutative: The permutation of multiplicands is admissible only when one of multiplicands is scalar or when the vector parts of multiplicands are proportional.

A conjugate quaternion of the given quaternion Q, designated as $\tilde{Q}$, is represented as

$$\tilde{Q} = q_1 - q.$$

The norm of a quaternion Q is given by

$$\|Q\| = Q\tilde{Q} = \tilde{Q}Q = q_1^2 + q_2^2 + q_3^2 + q_4^2.$$

Thus, the norm of a quaternion, a scalar $\|Q\| = 0$, implies that $q_1 = q_2 = q_3 = q_4 = 0$. When the norm is $\|Q\| = 1$, the quaternion is referred to as normalized.

The product of two quaternions is equal to zero only when one of the multiplicands is equal to zero.

If Q is not the zero quaternion, then

$$Q \frac{\tilde{Q}}{\|Q\|} = \frac{\tilde{Q}}{\|Q\|} Q = H_0.$$

The inverse of the given quaternion Q is the quaternion $Q^{-1} = \frac{\tilde{Q}}{\|Q\|}$, for which the following equality holds:

$$QQ^{-1} = Q^{-1}Q = H_0.$$

Consequently, the equation $S = QP$, when P is unknown, can be solved by multiplying its left- and right-hand sides by Q^{-1}, which yields

$$Q^{-1}QP = P = Q^{-1}S = \frac{\tilde{Q}S}{\|Q\|}.$$

The product of quaternions having zero scalar parts is represented as

$$qp = (-q_2p_2 - q_3p_3 - q_4p_4)H_0 + (q_3p_4 - q_4p_3)H_1 - (q_2p_4 - q_4p_2)H_2 + (q_2p_3 - q_3p_2)H_3.$$

Accordingly, the scalar and vector parts of a product are equal:

$$\text{scal}(qp) = -(q_2p_2 + q_3p_3 + q_4p_4)H_0,$$
$$\text{vect}(qp) = (q_3p_4 - q_4p_3)H_1 - (q_2p_4 - q_4p_2)H_2 + (q_2p_3 - q_3p_2)H_3.$$

The changing of places of multiplicands does not change a scalar part of the product, but changes the sign of the basis matrix H_3:

$$\text{scal}(qp) = \text{scal}(pq),$$
$$\text{vect}(qp) = -\text{vect}(pq).$$

2.4.2 Singular Value Decomposition

According to (2.6), the binding energy between FPs is determined by a bilinear form. It is also known (e.g., see Horn and Johnson, 1986) that extremal values of a bilinear form are determined by the so-called singular value decomposition (SVD) of a matrix. Let us consider this important notion in greater detail.

First, the relevant expressions could be simplified by adopting some concepts from linear algebra. A matrix A is called a symmetric matrix if $A = A^T$, and a Hermitian matrix if $A = A^*$.

A matrix A is called an orthogonal (unitary) matrix if $AA^T = E = AA^T$ for a real-valued matrix and $AA^* = A^*A = E$ for a complex-valued matrix.

A matrix A is called a normal matrix, if $AA^T = A^TA$ (for a real-valued matrix) or $A^*A = AA^*$ (for a complex-valued matrix).

Real symmetric and orthogonal matrices, and also Hermitian and unitary matrices, are normal.

The product of orthogonal (unitary) matrices is itself an orthogonal (unitary) matrix.

For an arbitrary matrix A of dimension $m \times n$ there exists a singular value decomposition, i.e., a representation

$$A = V\Sigma W^*, \qquad (2.8)$$

where V of dimension $m \times m$ and W of dimension $n \times n$ are unitary square matrices satisfying a unitary criterion:

$$VV^* = V^*V = E, \quad WW^* = W^*W = E.$$

The matrix V^* is defined for a matrix V via transposition and subsequent replacement of matrix elements by conjugates of the unit matrix E. The columns of matrices V and W are called respectively the left and right singular vectors of the matrix A. The matrix Σ consists of a diagonal square cell of dimension $r \times r$, where $r = \min\{m, n\}$, with nonnegative elements on the main diagonal, and if $n \neq m$ has additional zero rows or columns, then

$$\Sigma = \begin{cases} [\Sigma' \quad \varnothing] & \text{if } m < n, \\ \begin{bmatrix} \Sigma' \\ \cdots \\ 0 \end{bmatrix} & \text{if } m > n, \\ \Sigma' & \text{if } m = n, \end{cases}$$

where $\Sigma' = \mathrm{diag}(s_1, s_2, \ldots, s_r)$, $s_1 \geq s_2 \geq \cdots \geq s_r$.

The values $s_j, j=1,\ldots, r$, are known as singular values of the matrix A, and the representation $A = V\Sigma W^*$ is called the SVD of the matrix A. The singular values are unique to a matrix A.

Let us consider a composite matrix of dimension $2(m + n) \times 2(m + n)$,

$$A = \begin{bmatrix} 0 & A^* \\ A & 0 \end{bmatrix},$$

and its characteristic equation

$$\det[A - \lambda E_{n+m}] = \det\begin{bmatrix} \lambda E_n & A^* \\ A & -\lambda E_m \end{bmatrix} = 0.$$

Taking into account the SVD (2.8) of the matrix A and the unitary criterion, we can write

$$\det\begin{bmatrix} \lambda E_n & A^* \\ A & -\lambda E_m \end{bmatrix} = \det\left\{ \begin{bmatrix} V_n & 0 \\ 0 & W_m \end{bmatrix} \begin{bmatrix} -\lambda E_n & \Sigma^* \\ \Sigma & -\lambda E_m \end{bmatrix} \begin{bmatrix} V_n^* & 0 \\ 0 & W_m^* \end{bmatrix} \right\}.$$

Let us consider the equation

$$\det\begin{bmatrix} -\lambda E_n & \Sigma^* \\ \Sigma & -\lambda E_m \end{bmatrix} = 0$$

under the following conditions:

$n=m$

$$\det\begin{bmatrix} -\lambda E_n & \Sigma^* \\ \Sigma & -\lambda E_m \end{bmatrix} = (\lambda^2 - s_n^2)(\lambda^2 - s_{n-1}^2)\cdots(\lambda^2 - s_1^2);$$

$n>m$

$$\det\begin{bmatrix} -\lambda E_n & \Sigma^* \\ \Sigma & -\lambda E_m \end{bmatrix} = -\lambda^{m-n}(\lambda^2 - s_n^2)\cdots(\lambda^2 - s_1^2);$$

$n<m$

$$\det\begin{bmatrix} -\lambda E_n & \Sigma^* \\ \Sigma & -\lambda E_m \end{bmatrix} = (-\lambda)^{m-n}(\lambda^2 - s_m^2)(\lambda^2 - s_{m-1}^2)\cdots(\lambda^2 - s_1^2).$$

So, among the solutions of the equation $\det[A - \lambda E_{n+m}] = 0$ there are diagonal elements of a matrix Σ, and $(n - m)$ roots of this equation are equal to zero. Thus, the singular values $s_1, s_2, \ldots, s_r$, where $r = \min\{m, n\}$, constitute very important characteristic roots of the composite matrix A that uniquely define this matrix. Unfortunately, the unitary matrices V_n, V_m participating in the SVD (2.8) are ambiguously defined by the matrix A.

It is known (e.g., see Horn and Johnson, 1986) that the singular values of a matrix A do not change if this matrix is multiplied on the left and on the right by any unitary matrices U_L and U_R of appropriate dimensions ($m \times m$) and ($n \times n$).

If $A = V\Sigma W^*$ and $B = U_L A U_R$, then $B = \tilde{V}\Sigma \tilde{W}^*$, where the matrices $\tilde{V} = U_L V$, $\tilde{W} = U_R^* W$ are unitary. Therefore, there is an SVD of matrices A and B, in which the same matrix Σ takes part, containing the singular values of its diagonal terms.

The matrices A, A^*, A^T have identical singular values.

For the definition of singular values of a matrix A, we frequently use eigenvalues of the Hermitian matrices A^*A and AA^*. If the matrix A has dimension ($m \times n$), then the matrix A^*A has dimension ($n \times n$), and the matrix AA^* has dimension ($m \times m$).

From the equalities

$$A^*A = V\Sigma^*\Sigma V^*, \quad AA^* = W\Sigma\Sigma^* W^*,$$

$$\Sigma^*\Sigma = \begin{cases} \text{diag}\{s_m^2,\ldots,s_1^2,\ldots 0\}, & \text{if } m < n, \\ \text{diag}\{s_n^2, s_{n-1}^2,\ldots s_1^2\}, & \text{if } m \geq n, \end{cases}$$

$$\Sigma\Sigma^* = \begin{cases} \text{diag}\{s_m^2,\ldots,s_1^2\}, & \text{if } m \leq n, \\ \text{diag}\{s_n^2, s_{n-1}^2,\ldots s_1^2, 0\ldots 0\}, & \text{if } m > n, \end{cases}$$

it is obvious that

$$s_{r-j}(A) = \sqrt{\lambda_{n-j}(A^*A)} = \sqrt{\lambda_{m-j}(AA^*)}, \quad j = 0,1,\ldots,r-1,$$

where $r = \min\{m, n\}$.

If A is a square matrix, then matrices A^*A and AA^* also have identical eigenvalues equal to squared singular values of the matrix A.

The SVD of an arbitrary real-valued matrix A has the following appearance:

$$A = s_1 L_1 R_1^T + s_2 L_2 R_2^T + \cdots + s_r L_r R_r^T, \tag{2.9}$$

where
  s_i are singular values of matrix A,
  L_i, R_i are its left and right singular vectors and,
  r is the rank of the matrix.

The singular values and singular vectors satisfy the following relations:

$$s_1 \geq s_2 \geq \cdots \geq s_r \geq 0, \quad s_i = L_i^T A R_i, \quad L_i^T L_i = 1, \quad R_i^T R_i = 1, \quad i = 1,\ldots,r.$$

It is known that the SVD possesses very useful properties for theory and applications (Horn and Johnson, 1986). Specifically, every matrix over the field of real numbers has real singular values and vectors. Moreover, the SVD is stable to small disturbances of the matrix, i.e., the SVD of every matrix is a well-conditioned problem. The SVD can generally be computed according to a rather simple and reliable scheme:

$$\left. \begin{array}{c} R_{(k+1)}^T = L_{(k)}^T A, \ R_{(k+1)} = \dfrac{R_{(k+1)}}{|R_{(k+1)}|}, \\[2mm] L_{(k+1)} = A R_{(k+1)}, \ L_{(k+1)} = \dfrac{L_{(k+1)}}{|L_{(k+1)}|}, \end{array} \right\} \quad (2.10)$$

$$s_k = L_k^T A R_k, \ |s_{k+1} - s_k| \le \varepsilon,$$

where $k=0,1,2,\ldots$ is the number of the iteration, and ε is the given precision of calculation. It can be shown that for arbitrary unit vectors $L_{(0)}$, $R_{(0)}$, iterations by scheme (2.10) generally converge to the singular vectors L, R, corresponding to the maximal singular value $s^* = L^T A R$.

It is worth noting that such properties are beyond the spectral value decomposition (Van der Waerden, 1957), which actually forms the basis for multivariate statistical analysis. Unlike the SVD, eigenvalues and eigenvectors of the spectral value decomposition are real only for symmetric real matrices; generally nonsymmetric real matrices possess a complex spectrum, and obtaining it is not so simple (Gantmakher, 1988). Besides, a matrix can be ill-conditioned concerning computation of its eigenvalues, which creates well-known problems (Wilkinson, 1965).

The SVD representation (2.9) can be realized by the *deflation method*. For this purpose the maximal singular value s_1 of the matrix $A_{(1)} = A$ and corresponding singular vectors L_1, R_1 are determined by the iterative scheme (2.10). Then the matrix

$$A_{(2)} = A_{(1)} - s_1 L_1 R_1^T$$

is formed, and its maximal singular value s_2 and corresponding singular vectors L_2, R_2 are determined by the scheme (2.10). In general, at every iterative step number p, the matrix

$$A_{(p)} = A_{(p-1)} - s_{p-1} L_{p-1} R_{p-1}^T$$

is formed, and its maximal singular value s_p and corresponding singular vectors L_p, R_p are determined by the scheme (2.10).

2.4.3 The SVD of Interval Matrices

As a rule, practical applications require addressing the issue of parametric uncertainty of complex processes. The parametric uncertainty can be characterized by a relationship of the true values of the parameters of the process to intervals with known boundaries. In fact, the interval model of a system reflects a realistic situation with the information of the values of its parameters, when the boundaries of such intervals are known a priori. Therefore, mathematical models of such objects can be represented by using the rules and nomenclature of interval mathematics. Operating in the interval space has the following mathematical specifics:

- an incomplete algebraic structure;
- an incomplete ordinal structure;
- lack of the property of rigorous distributivity.

Due to such specifics, the majority of the tasks of interval analysis are NP-complete.

The results of the development of an approach to SVD of interval matrices are represented below.

Consider a real valued interval matrix

$$[A] \in M_{m,m}(I(R)),$$

where

$$[A] = [a_{ij}]_{i,j=1}^{m}, \quad [a_{ij}] = [\underline{a}_{ij}, \overline{a}_{ij}],$$

$\underline{a}_{ij}, \overline{a}_{ij}$ are the lower and upper boundaries of the intervals, and $M_{m,m}(I(R))$ is a set of $m \times m$ matrices whose elements are real-valued intervals $[\underline{a}, \overline{a}] = (a \in R \wedge \underline{a} \leq a \leq \overline{a})$.

By the interval matrix $[A]$ we shall understand the set of real-valued matrices of dimension $m \times m$

$$\{A \mid A \in [A]\}.$$

Let $\phi(x): IR^m \to IR^m$ be an operator of multiplication by a thin matrix; i.e.,

$$\phi([x]) = G[x]$$

for some $G \in R^{m \times m}$, $G = (g_{ij})$.

For the mapping considered herein we shall take advantage of the standard procedure of embedding the interval space IR^m in the usual well-investigated Euclidean space R^{2m}. The standard embedding $\sigma : [IR]^m \to IR^{2m}$ works as follows:

$$([x]_1, [x]_2, \ldots, [x]_m) \mapsto (-\underline{x}_1, -\underline{x}_2, \ldots, -\underline{x}_m, \overline{x}_1, \overline{x}_2, \ldots, \overline{x}_m),$$

i.e., when the first, second, ..., mth the components of the vector $\sigma([x])$ rely on the low limits of the elements $x_1, x_2, \ldots, x_m$, taken with the opposite sign, and the $(m+1)$th, ..., $2m$th components of $\sigma([x])$ rely on the upper limits of $x_1, x_2, \ldots, x_m$, respectively.

The standard embedding represented by σ is the map that according to (Shary, 1995),

1. reflects the additive algebraic structure IR^m, i.e.,

$$\sigma([x]+[y]) = \sigma([x]) + \sigma([y]) \text{ for any } [x] \text{ and } [y],$$

2. reflects the topological structure IR^m; i.e., both the map $\sigma : IR^m \to R^{2m}$ and its inverse $\sigma^{-1} : R^{2m} \to IR^m$ are continuous.

The standard embedding is equivalent to a convolution of the interval space IR^m with the system component space R^{2m}.

Using the standard embedding $\sigma : IR^m \to R^{2m}$, we see that the induced map $\sigma \cdot \phi \cdot \sigma^{-1}$ is a linear transformation of the space R^{2m}.

For the standard embedding σ, the block matrix representing the linear transformation $\sigma \cdot \phi \cdot \sigma^{-1}$ is of dimension $2m \times 2m$:

$$G^\sigma = \begin{vmatrix} G^+ & G^- \\ G^- & G^+ \end{vmatrix},$$

where the matrices $G^+ = (g_{ij}^+)$ and $G^- = (g_{ij}^-)$ are the positive and negative parts; i.e.,

$$g_{ij}^- = \max\{-g_{ij}, 0\}, \quad g_{ij}^+ = \max\{g_{ij}, 0\}.$$

The matrix G^σ is known as a companion of the matrix G. To realize the SVD for the companion matrix G^σ, we applied the efficient scheme (2.10) and the reduction method.

The algorithm of SVD of an accompanying matrix G^σ consists of the following steps:

1. For the maximum singular value of a matrix G^σ, the right and left singular vectors are defined by the scheme (2.10): $s_{1\max}, L_1, R_1$.
2. The auxiliary matrix G^σ_1 is formed:

$$G_1^\sigma = G^\sigma - s_{1\max} L_1 R_1^T.$$

3. For the maximum singular value of a matrix G_1^σ, the right and left singular vectors are defined: $s_{2\max}, L_2, R_2$.
4. The auxiliary matrix G_2^σ is formed:

$$G_2^\sigma = G_1^\sigma - s_{2\max} L_2 R_2^T.$$

5. For the maximum singular value of the matrix G_2^σ, the right and left singular vectors are defined: $s_{3\max}, L_3, R_3$.

The algorithm for the definition of the spectral radius of a symmetric interval matrix is established below.

Let the interval matrix $[A]$ be given. Let us also assume that the given interval matrix $[A]$ is symmetric, i.e., the symmetric interval matrix is defined as the set of matrices of an aspect

$$[A]^{\text{sym}} = \left[\underline{A}^{\text{sym}}, \overline{A}^{\text{sym}}\right] = \left\{A \in R^{m \times m} \,\middle|\, A = A^T, \underline{A}^{\text{sym}} \leq A \leq \overline{A}^{\text{sym}}\right\},$$

where

$$\underline{A}^{\text{sym}}, \overline{A}^{\text{sym}} \in R^{m \times m}, \underline{A}^{\text{sym}} = \left(\underline{A}^{\text{sym}}\right)^T, \overline{A}^{\text{sym}} = \left(\overline{A}^{\text{sym}}\right)^T.$$

To label a symmetric matrix, the symbol $[A]^{\text{sym}} = \left([A]^{\text{sym}}\right)^T$ is used. For an interval matrix $[A]$ we shall define a maximum eigenvalue as

$$\lambda([A]) = \max \{\text{Re } \lambda, \lambda \text{ an eigenvalue of any matrix } A \in [A]\}.$$

If an interval matrix [A] is symmetric, then the maximum eigenvalue is represented as

$$\lambda([A]) = \max \{\lambda_{\max}(A);\ A \text{ symmetric and } A \in [A]\}.$$

It has been shown (Alefeld, 1983) that the procedure for eigenvalue definition of an interval matrix [A] is NP-complete; i.e., the set of test matrices increases exponentially with the dimensions of the initial matrix.

The algorithm of the definition of the spectral radius of a symmetric interval matrix allows us to adapt the aforementioned algorithms of the evaluation of eigenvalues and singular values and to reduce the computational complexity through purposely constructed sets of vectors and matrices.

The algorithm of the definition of the spectral radius of the symmetric interval matrix [A] consists of the following steps:

1. A set Z is formed:

$$Z = \{z \in R^m;\ z_j \in \{-1, 1\}\ \forall\, j\}.$$

2. Create a set of matrices A_z, $z \in R^m$; each component of the matrix A_z is represented as

$$(A_z)_{ij} = \begin{cases} \overline{a_{ij}}, & \text{if } z_i z_j = 1, \\ \underline{a_{ij}}, & \text{if } z_i z_j = -1, \end{cases}$$

$$i, j = 1, \ldots, m.$$

The above analysis indicates that for any $z \in Z$ the constructed matrices A_z are symmetric, $A_z \in [A]$.

3. As in (2.10), the spectrum of the eigenvalues of the matrices A_z is defined: $\lambda_k(A_z)$, $k=1,\ldots,m$, $z \in R^m$, and a maximum eigenvalue is selected: $\overline{\lambda_z} = \max\limits_{k} \lambda_k(A_z)$.

4. A maximum eigenvalue is defined:

$$\lambda([A]) = \left\{\max\limits_{k} \overline{\lambda_z}(A_z)\right\}.$$

3
Pattern Recognition

3.1 A Mathematical Model of Molecular Recognition

For solving problems of pattern recognition, modern AI widely uses an approach based on ANN. Such networks represent computational systems of parallel distributed processing, also called *neurocomputers* (Wasserman, 1990). The main feature of ANN is the ability to learn a processing rule, without its being explicitly assigned, by following patterns like "input–output" and "situation–action."

However, there is another, possibly more promising, biological model. Modern molecular biology has discovered that remarkably elegant, precise, and reliable mechanisms of recognition are utilized by proteins, and that such mechanisms apparently play a key role in both immune and intellectual processes (see Section 1.2). These mechanisms are also parallel and distributed, like ANN.

Accordingly, this chapter develops an approach to pattern recognition by immunocomputing. The mathematical basis of the approach shows a special kind of interaction between FPs that allows us to represent the main operations of linear algebra.

3.1.1 The Representation of a Bilinear Form

Consider an FP with n links (let us call such an FP an *n-peptide*). Let each link be of the kind $Q_C(\varphi)$ (see Example 2.2). Such links can be represented by complex numbers, which are special kinds of quaternions:

$$Q_C(\varphi) = H_0 \cos\varphi + H_1 \sin\varphi.$$

To every such n-peptide with links $Q_C(\varphi_i)$, $i=1,\dots,n$, assign a unit column vector $X^{\langle n+1 \rangle}$ of dimension $n+1$ by the following recurrent rules:

$$X^{\langle 1\rangle}=[1],\quad X^{\langle i+1\rangle}=\left[X^{\langle i\rangle}\cos\varphi_i,\sin\varphi_i\right]^T, i=1,\ldots,n. \qquad (3.1)$$

For example,

$$X^{\langle 2\rangle}=[\cos\varphi_1,\sin\varphi_1]^T,\ X^{\langle 3\rangle}=[\cos\varphi_1\cos\varphi_2,\sin\varphi_1\cos\varphi_2,\sin\varphi_2]^T,\ldots.$$

Such a vector $X^{\langle n+1\rangle}$ can be assigned to any n-peptide. And conversely, to any unit vector

$$X^{\langle n+1\rangle}=[x_1,\ldots,x_{n+1}]^T,\ X^T X=1,$$

the corresponding n-peptide can be uniquely assigned. The torsion angles of such an FP $\varphi_1,\ldots,\varphi_n$ can be defined, for example, by the following rules:

$$\varphi_n=\arcsin(x_{n+1}),\ \varphi_{i-1}=\arcsin\left(\frac{x_i}{\prod_{k=i}^{n}\cos\varphi_k}\right),\ -\frac{\pi}{2}\leq\varphi_i\leq\frac{\pi}{2},\ i=n,n-1,\ldots,2;\ \text{if}$$

$x_1<0\ \ \varphi_1=\pi-\varphi_1$.

For such FPs, let us consider a special kind of binding.

Definition 3.1 *Homologous binding* is a binding of two n-peptides such that their links with equal numbers bind sequentially, and any component of their energy matrices and/or binding matrix can accept the value of the free energy or binding energy of the previous links.

Consider a homologous binding of two n-peptides, where all their links bind: $Q_c(\varphi_i)$, $Q_c(\psi_i)$, $i=1,\ldots,n$. Figure 3.1 shows a scheme for such a complete binding, where horizontal lines denote links of the same FP, vertical dotted lines denote binding links, arrows denote the consequence of bindings, and $w^{(i)}$ denotes the binding energy between the corresponding links number i.

$$Q_c(\varphi_1)\ -\ Q_c(\varphi_2)\ -\ \cdots\ -\ Q_c(\varphi_n)$$
$$\vdots\ w^{(1)}\rightarrow\ \vdots\ w^{(2)}\rightarrow\ \cdots\ \rightarrow\ \vdots\ w^{(n)}$$
$$Q_c(\psi_1)\ -\ Q_c(\psi_2)\ -\ \cdots\ -\ Q_c(\psi_n)$$

Figure 3.1 Complete homologous binding.

3.1 A Mathematical Model of Molecular Recognition

Let the binding matrices of the links have the following appearance:

$$W^{\langle 1 \rangle} = \begin{bmatrix} 1 & 0 \\ 0 & 1 \end{bmatrix}, \quad W^{\langle i+1 \rangle} = \begin{bmatrix} w^{(i)} & 0 \\ 0 & 1 \end{bmatrix}, \quad i=1,\ldots,n-1.$$

Then the binding energy of the links can be represented in vector–matrix notation:

$$w^{(i)} = -[Q_c(\varphi_i)]^T W^{\langle i \rangle} [Q_c(\psi_i)],$$

where $i=1,\ldots,n$. Hence, in accordance with (3.1), derive

$$w^{(i)} = -w^{(i-1)} \cos\varphi_i \cos\psi_i - \sin\varphi_i \sin\psi_i. \qquad (3.2)$$

Let unit vectors $X^{\langle i \rangle}$, $Y^{\langle i \rangle}$ of type (3.1) represent two FPs. Then their scalar product is equal to

$$[X^{\langle i+1 \rangle}]^T [Y^{\langle i+1 \rangle}] = [X^{\langle i \rangle}]^T [Y^{\langle i \rangle}] \cos\varphi_i \cos\psi_i + \sin\varphi_i \sin\psi_i. \qquad (3.3)$$

A comparison of (3.3) and (3.2) proves the following proposition:

Proposition 3.1 The scalar product of unit vectors can be represented by the (negative) binding energy of the complete homologous binding of two FPs.

Consider now an incomplete (partial) homologous binding of links $Q_c(\varphi_i) \leftrightarrow Q_c(\psi_i)$ of two FPs, which are shown in Figure 3.2. In this case, between any links of type $Q_c(\varphi_i)$ and $Q_c(\varphi_{i+1})$ there exists a *secondary structure (loop)* of type $Q_X^{(i)}(\varphi_1) - \cdots - Q_X^{(i)}(\varphi_i)$, where i is a number of the loop. Links of the loops have the following appearance:

$$Q_X(\varphi) = H_0 \cos\frac{\varphi}{2} + H_1 \sin\frac{\varphi}{2}.$$

Figure 3.2 shows the free energy of every loop as v with a corresponding index.

From the above we have the following proposition:

Theorem 3.1 The values of an arbitrary bilinear form over unit vectors can be represented as the values of the binding energy (with the sign "minus") of the homologous binding of two FPs.

The proof of this theorem can be found in (Tarakanov, 1999a). It should be mentioned, though, that it includes the definition of such energy matrices and binding matrices of two FPs for an arbitrary matrix $A=[a_{ij}]$, which gives the following recursive equation for the binding energy:

$$w^{(n)} = -[Q_d(\varphi_n)]^T W^{(n-1)} [Q_d(\psi_n)],$$

which coincides with the bilinear form

$$w^{(n)} = -[X^{\langle n+1 \rangle}]^T A [Y^{\langle n+1 \rangle}]. \quad (3.4)$$

This result means that every matrix can be treated as a mode of assigning the binding energy of the homologous binding of FPs.

$$\begin{array}{ccccc}
v_{\varphi_1}^{(1)} & & v_{\varphi_{n-1}}^{(1)} & & v_{\varphi_{n-1}}^{(n-1)} \\
Q_d(\varphi_1) - Q_X^{(1)}(\varphi_1) - Q_d(\varphi_2) - \cdots - Q_X^{(1)}(\varphi_{n-1}) - \cdots - Q_X^{(n-1)}(\varphi_{n-1}) - Q_d(\varphi_n) \\
\Big| w^{(1)} \rightarrow \quad \Big| w^{(2)} \rightarrow \quad \cdots \rightarrow \quad \Big| w^{(n)} \\
Q_d(\psi_1) - Q_X^{(1)}(\psi_1) - Q_d(\psi_2) - \cdots - Q_X^{(1)}(\psi_{n-1}) - \cdots - Q_X^{(n-1)}(\psi_{n-1}) - Q_d(\psi_n) \\
v_{\psi_1}^{(1)} & & v_{\psi_{n-1}}^{(1)} & & v_{\psi_{n-1}}^{(n-1)}
\end{array}$$

Figure 3.2 Partial homologous binding.

3.1.2 Recognition between FPs

Definition 3.2 Define the *recognition* between two FPs as their binding.

In other words, two FPs recognize one another if they interact with a binding energy less than or equal to the value of the threshold (see Definition 2.5). Thus we can say that the lower the binding energy, the better the recognition.

Let us now have some set of FPs and a mode of determining the binding energy for every pair of FPs. For example, we can consider the matrix of binding existing for any pair of FPs. Then we can define a method of computation of the matrix of binding that is dependent on the states (angles) of the interacting FPs. Then matrices of binding will generally be different for different pairs of FPs. In any case, the following problem is worth considering: For a given set of FPs and a given method of determining the binding energy, find those pairs of FPs that can recognize one another. A special case of this problem is detecting the pairs of FPs that allow for best recognition (those in which the binding energy is minimal).

Let us demonstrate that in the case of homologous binding the above problem has a complete theoretical solution.

Actually, in the case of complete homologous binding (Figure 3.1), the binding energy is the scalar product of the unit vectors (3.3) corresponding to the pair of FPs, and geometric considerations make it obvious that the lower the angle between vectors (in n-dimensional space), the better the recognition. Hence, the best recognition in general occurs between FPs with coinciding vectors. This case corresponds to identical FPs with equal torsion angles $\varphi_k = \psi_k$, $k=1,\ldots,n$.

Now consider a general case of homologous binding (Figure 3.2). According to Theorem 3.1, any matrix can be treated as the means of assigning the binding energy to FPs, numbers of links of which correspond to the dimension of the matrix. Let such a matrix A be given and the task be to determine the FPs with the best recognition.

Formally, this task can be reduced to minimizing the Lagrange function f over vectors X, Y:

$$\min_{X,Y} f : f = -X^T A Y + \lambda_1 (X^T X - 1) + \lambda_2 (Y^T Y - 1),$$

where λ_1, λ_2 are Lagrange multipliers. According to the rules of vector–matrix differentiation (Rao, 1967), we obtain the derivatives of the function f with respect to vectors X, Y as the vectors

$$[f'_X] = 2\lambda_1 X - AY,$$
$$[f'_Y] = 2\lambda_2 Y - A^T X.$$

These derivatives have to be equal to zero if the vectors minimize the function. In this case, multiplying the left sides of the first equation by X^T and the second equation by Y^T, we obtain $2\lambda_1 = 2\lambda_2 = X^T A Y$. Define $s^* = 2\lambda_1 = 2\lambda_2$. Then the best recognition is reached by FPs whose vectors satisfy the equations

$$AY = s^* X,$$
$$X^T A = s^* Y^T.$$
(3.5)

But these equations also determine singular vectors of the matrix A, corresponding to its singular value s^* (Horn and Johnson, 1986). Hence, the best recognition (the minimal binding energy) is achieved with the pair of FPs whose torsion angles form singular vectors corresponding to the maximal singular value of the matrix A. Moreover, the singular value with the sign "minus" gives the minimum binding energy over all pairs of FPs.

Therefore, homologous binding reduces the task of determining the best recognizing FPs to determining the maximal singular value and corresponding singular vectors of the matrix that define the binding energy. The following iterative scheme gives such vectors in practice:

$$Y_{(k+1)}^T = X_{(k)}^T A, \quad Y_{(k+1)} = \frac{Y_{(k+1)}}{|Y_{(k+1)}|},$$
$$X_{(k+1)} = A Y_{(k+1)}, \quad X_{(k+1)} = \frac{X_{(k+1)}}{|X_{(k+1)}|}, \quad (3.6)$$

where $k=0,1,2,...$ is the number of the iteration.

It can be shown that for arbitrary unit vectors $X_{(0)}$, $Y_{(0)}$), iterations by scheme (3.6) generally converge to the singular vectors X,Y, corresponding to the maximal singular value $s^* = X^T A Y$.

Therefore, recognition between FPs is reduced to a singular value decomposition (SVD) of an arbitrary matrix A, which determines the binding energy between the FPs. An important property of the SVD is its stability with respect to small disturbances of the matrix A. In terms of the model of FP considered, the recognition is stable under small disturbances of the parameters that determine the binding energy between FPs.

3.1.3 Specificity of Recognition

Let a set $\{Z\}$ be given, consisting of $n=mr$ numbers

$$\{z_{11},...,z_{1r},...,z_{m1},...,z_{mr}\}.$$

According to Theorem 3.1, any such set can be treated as a set of coefficients that determine the binding energy between FPs. Specifically, consider this set as a column vector of dimension $mr \times 1$:

$$Z=[z_{11},...,z_{mr}]^T.$$

Normalize this vector: $Z=\mu Z_e$, where $\mu = |Z|$, $Z_e^T Z_e = 1$. Then the best recognition is obtained by the pair of FPs with vectors $X=Z_e$, $Y=[1]$, and the binding energy is $w=-\mu$. In fact, these vectors and their binding energy (with the sign "minus") determine the SVD of the initial set of numbers, represented as the vector $Z=\mu X Y^T$.

The same set of numbers can also be treated as a matrix of dimension $m \times r$:

$$A_Z = \begin{bmatrix} z_{11} & \cdots & z_{1r} \\ \cdots & \cdots & \cdots \\ z_{m1} & \cdots & z_{mr} \end{bmatrix},$$

which has the SVD

$$A_Z = s_1 X_1 Y_1^T + \cdots + s_r X_r Y_r^T,$$

where X_i, Y_i, s_i are singular vectors and singular values of the matrix A_Z, $i=1,...,r$, $r \leq m$, and r is the rank of the matrix.

The matrix A_Z, as well as the vector Z, can also determine the binding energy, but between pairs of FPs with different numbers of links. In this context of recognition of FPs, one can consider the concept of *folding a vector to a matrix*. Let us discuss a *set of standard pairs of FPs* and *standard binding energies*, by which we understand singular vectors and singular values (with the sign "minus") connected with a vector or matrix representation of the initial set of numbers $\{Z\}$.

Specifically, for a vector representation we have a standard pair of FPs $\{Z_e, [1]\}$ with a standard binding energy $-\mu$. For the matrix representation we have a set of standard pairs $\{X_i, Y_i,\}$ with binding energies $-s_i$, where $i=1,...,r$, and r is the rank of the matrix.

Now let the binding energy be determined by another set of numbers, the capacity of which is equal to the capacity of the initial set: $n=mr$. Obviously, the binding energies of the standard pairs can differ from their standard values. Then introduce the quantitative measure of such deviation.

Definition 3.3 *Specificity of recognition* (called for convenience *specificity*) is the sum of the squares of deviations of the binding energies from their standard values over all the standard pairs of FPs.

The introduction of such a measure of similarity between two sets of numbers that define the binding energy leads to the following result:

Theorem 3.2 Folding a vector to a matrix does not decrease the specificity.

Proof. Consider a disturbance $\{dZ\}=\{dz_{11},...,dz_{mr}\}$ of the initial set $\{Z\}$. Obtain the specificity measure ρ_Z for the pair $\{Z_e, [1]\}$ in the case of the vector representation of the disturbed binding energy. In this case the new binding energy for the standard pair is equal to

$$Z_e^T(Z+dZ) = Z_e^T Z + Z_e^T dZ = \mu + e_{11} dz_{11} + \cdots + e_{mr} dz_{mr} = \mu + d\rho, \quad (3.7)$$

where $Z_e=[e_{11},...,e_{mr}]^T$.

Now obtain the specificity ρ_A for the standard pairs $\{X_k, Y_k\}$, $k=1,\ldots,r$, in the case of the matrix representation of the disturbed binding energy.

Consider the SVD of the matrix A_Z:

$$A_Z = s_1 X_1 Y_1^T + \cdots + s_r X_r Y_r^T,$$

$$A_Z = s_1 \begin{bmatrix} x_{11} \\ \cdots \\ x_{1m} \end{bmatrix} [y_{11} \ \cdots \ y_{1r}] + \cdots + s_r \begin{bmatrix} x_{r1} \\ \cdots \\ x_{rm} \end{bmatrix} [y_{r1} \ \cdots \ y_{rr}].$$

Considering the vector representation $Z=\mu Z_e$ together with the SVD of the matrix A_Z, any element of the initial set $\{Z\}$ can be written as

$$z_{ij} = \mu e_{ij} = s_1 x_{1i} y_{1j} + s_2 x_{2i} y_{2j} + \cdots + s_r x_{ri} y_{rj}, \qquad (3.8)$$

where $i=1,\ldots,m$, $j=1,\ldots,r$.

In the case of matrix representation, the new binding energy for the first standard pair is

$$\rho_1 = X_1^T (A_Z + dA_Z) Y_1 = s_1 + X_1^T dA_Z Y_1 = s_1 + \sum_{i,j} x_{1i} y_{1j} dz_{ij} = s_1 + d\rho_1.$$

Analogously, for the kth standard pair, where $k=1,\ldots,r$, the new binding energy is equal to

$$\rho_k = s_k + d\rho_k, \quad d\rho_k = \sum_{i,j} x_{ki} y_{kj} dz_{ij}, \quad k = 1,\ldots,r.$$

The obtained values $d\rho, d\rho_k$ allow us to determine the specificity ρ_Z, ρ_A for the vector and matrix representations:

$$\rho_Z = (d\rho)^2, \quad \rho_M = (d\rho_1)^2 + \cdots + (d\rho_r)^2 = t_1^2 + \cdots + t_r^2. \qquad (3.9)$$

Express a value $d\rho$ for the vector representation by the values $d\rho_k$ of the matrix representation. From the formulas (3.7), (3.8) derive

$$d\rho = \sum_{i,j} e_{ij} dz_{ij},$$

$$d\rho = \sum_{k=1}^{r} \frac{s_k}{\mu} x_{k1} y_{k1} + \cdots + \sum_{k=1}^{r} \frac{s_k}{\mu} x_{ki} y_{kj} + \cdots + \sum_{k=1}^{r} \frac{s_k}{\mu} x_{km} y_{kr} = \sum_{k=1}^{r} \frac{s_k}{\mu} d\rho_k. \qquad (3.10)$$

Now demonstrate that $\rho_M \geq \rho_Z$.

According to formulas (3.9), (3.10), this inequality is equivalent to the inequality

3.1 A Mathematical Model of Molecular Recognition

$$t_1^2 + \cdots + t_r^2 \geq \left(\frac{s_1}{\mu}t_1 + \cdots + \frac{s_r}{\mu}t_r\right)^2.$$

Since the sum of the squares of elements of any matrix is equal to the sum of the squares of the singular values of the matrix

$$\mu^2 = s_1^2 + \cdots + s_r^2,$$

the inequality is reduced to the following equivalent inequality:

$$t_1^2(\mu^2 - s_1^2) + \cdots + t_r^2(\mu^2 - s_r^2) \geq 2\sum_{i<j} s_i s_j t_i t_j.$$

Transforming differences in the brackets to the form

$$\mu^2 - s_1^2 = s_2^2 + \cdots + s_r^2,$$
$$\mu^2 - s_k^2 = s_1^2 + \cdots + s_{k-1}^2 + s_{k+1}^2 + \cdots + s_r^2,$$
$$\mu^2 - s_r^2 = s_1^2 + \cdots + s_{r-1}^2,$$

we reduce the inequality to the following obvious expression:

$$\sum_{i<j}(s_i t_j - s_j t_i)^2 \geq 0. \qquad (3.11)$$

Hence, the equivalent expression $\rho_A \geq \rho_Z$ is correct, and it means that specificity either increases or remains the same in the case of folding a vector to a matrix.

This proof also implies that the procedure of folding a vector to a matrix increases the specificity measure, except in some rare cases, when inequality (3.11) becomes an equality. The first such case takes place when the rank of the matrix A_Z is equal to 1; i.e., all singular values, except the first one, are equal to zero: $s_2 = \cdots = s_r = 0$.

Suppose that the rank of the matrix A_Z is greater than one ($r > 1$), and consider the second case, when inequality (3.11) becomes an equality. This case takes place if and only if the conditions

$$s_i X_j^T (dA_Z) Y_j = s_j X_i^T (dA_Z) Y_i$$

are satisfied for every $i, j = 1, \ldots, r$, $i \neq j$. This means that the perturbed matrix dA_Z and the standard matrix A_Z have the same singular vectors and that all their singular values differ by only a constant factor. Hence, in this case the perturbed matrix ought to differ from the standard matrix only by some constant factor c:

$$[dA_Z] = c[A_Z].$$

Furthermore, inequality (3.11) shows that except in the above two cases there are no other situations in which the left side of (3.11) is equal to zero. Hence, the following proposition has been proved:

Proposition 3.2: If the rank of the standard matrix is greater than 1, and the perturbed matrix is not equal to the standard matrix multiplied by some constant, then folding a vector to a matrix definitely increases the specificity of recognition.

3.2 Pattern Recognition by Immunocomputing

3.2.1 The General Task of Pattern Recognition

Pattern recognition can be defined as follows:
Let us treat real values $x_1, ..., x_n$ as a set of characteristics (or *indicators*). Consider an arbitrary vector $X=[\ x_1,\ ...,\ x_n]^T$ as a pattern that belongs to an *indicator space* $\{X\}$. Consider that the space can be partitioned into subsets (classes) $\{X\}_k$, $k=1, 2, ..., k_n$. Then recognition of X implies the determination of a class k such that $X \subset \{X\}_k$, and learning implies the partitioning (classification) of the indicator space. If the space is being partitioned into known classes (e.g., by experts), then the problem is defined as *supervised learning*. If the number of the classes k_n and the classes themselves are unknown a priori, then it is called *unsupervised learning*.

Classification of several vectors into one class can be viewed as aggregation, or generalization, of the initial data. These two actions with their opposite objectives, unifying similar vectors and separating distinctive vectors, are typical of classification. And such actions have to be formalized.

Therefore, the key feature of pattern recognition is evaluation of the degree of the similarity between two vectors. Mathematically, it is defined by the vector norm, which formalizes the notion of geometric proximity (distance):

$$\text{dist}(X_a, X_b) = \|X_a - X_b\|.$$

The following norms are the most widespread in pattern recognition:
The Euclidean norm

$$\|X_a - X_b\|_E = \left(\sum_i (x_{ai} - x_{bi})^2\right)^{1/2};$$

The Manhattan norm (also called *metric of districts*)

$$\|X_a - X_b\|_M = \sum_i |x_{ai} - x_{bi}|;$$

The Chebyshev norm

$$\|X_a - X_b\|_C = \max_i |x_{ai} - x_{bi}|.$$

Apart from vector norms, M. Candle (Faure, 1985) has proposed a convenient distance function: For every pair of indicators x_{al} x_{am} of numbers $l<m$ devoted to every ath vector of indicators

$$X_a^T = [x_{a1}\ x_{a2}\ ...\ x_{an}],$$

define a coefficient of comparison

$$\Delta_{lm}^a = \begin{cases} +1 \text{ if } x_{al} > x_{am}, \\ -1 \text{ if } x_{al} < x_{am}, \\ 0 \text{ if } x_{al} = x_{am}. \end{cases}$$

By this coefficient the Candle distance between two vectors X_a, X_b is given by the formula

$$\text{dist}(X_a, X_b) = 1 - \frac{2}{n^2 - n} \sum_{l<m} \Delta_{lm}^a \Delta_{lm}^b.$$

Moreover, if the components of both vectors have a totally coinciding order, then min(dist)=0; if the components have a totally opposite order, then max(dist)=2.

Using a vector norm, we can define a notion of distance between a vector and a class,

$$\text{dist}(\widetilde{X}, z_k) = \min_X \{\text{dist}(\widetilde{X}, X):\ X \subset \{X\}_k\},$$

as well as a notion of distance between classes,

$$\text{dist}(z_{k1}, z_{k2}) = \min_{X,Y} \{d(X,Y):\ X \subset \{X\}_{k1}, Y \subset \{X\}_{k2}\}.$$

From the point of view of pattern recognition, it is natural to suppose that the lower the distance, the bigger the similarity. Furthermore, the degree of similarity can be evaluated by a variable that is the inverse of the distance.

Theoretically, the result of learning has to be a partition of the indicator space $\{X\}$ on classes such that

$$\bigcup_{j=1}^{k}\{X\}_j = \{X\}, \{X\}_i \cap \{X\}_j = \varnothing, \forall i \neq j,$$

where $\varnothing$ is the empty set. This property is called *separability* (ability to be partitioned), and it means (1) that classes cover all the indicator space and (2) that classes do not intersect. In addition, the requirement of *compactness* has to be observed, which implies that indicators of the same class are closer to one another than to indicators of any other class. However, both these properties are rare in practice, because they have a strong dependence upon the successful selection of indicators.

When the learning problem is solved and the indicator space is partitioned into the classes, then the recognition problem has to be solved. Usually, the recognition procedure is reduced to obtaining for every class $\{X\}_k$ a *solving rule*, such as

$$\widetilde{X} \in \{X\}_k : f_k(\widetilde{X}) > f_i(\widetilde{X}), i=1,\ldots,k_c, i \neq j,$$

where $f_k(X)$ is a *solving function*, and k_c is the general number of the classes.

If it is considered that the solving function has a deterministic nature, then its construction can be reduced to an approximation task, with decomposition by basic functions. If a solving function has a statistical nature, then the most widespread approach is based on the *Bayesian rule*

$$p(\{X\}_k/\widetilde{X})\, p(\widetilde{X}) = p(\widetilde{X}/\{X\}_k)\, p(\{X\}_k),$$

where $p(\{X\}_k)$ is the probability of the class $\{X\}_k$, and $p(\widetilde{X}/\{X\}_k)$ is a conditional probability of the vector of indicators such that the class $\{X\}_k$ is known. This conditional probability is called the *likelihood function*, and it can be computed or obtained by observation. Moreover, as a solving function for the class k we can use the conditional probability $p(\{X\}_k/\widetilde{X})$ of this class attached to the given vector of indicators $\widetilde{X}$, and this function is determined from the Bayesian rule. Usually, by these functions a *risk function* is assigned to every variant of recognition, and the vector is attached to the class where risk is minimal.

In general, pattern recognition represents an actively developing direction of AI (see Tarakanov et al., 1993). As previously noted (Faure, 1985), many methods of classification and recognition have been proposed, where this number, figuratively speaking, is equal to the number of the concrete tasks plus the number of researchers solving these tasks.

Nevertheless, the evolving approach to pattern recognition by IC shows some difference from other approaches. The main feature of this approach consists in treating an arbitrary pattern as a way of setting a binding energy for a bilinear form (3.4). Mathematically, this approach is based on the properties of the SVD of an arbitrary matrix over the field of real numbers. According to this approach, the task of pattern recognition can be solved as follows.

3.2.2 Supervised Learning

Folding Vectors to Matrices

Fold a vector X of dimension $n \times 1$ to a matrix A of dimension $n_P \times n_R = n$. According to Theorem 3.2, this operation increases the specificity of recognition.

Learning

Form matrices $A_1, \ldots, A_k$ for all classes $c=1,\ldots,k$ and compute their singular vectors:

$$\{P_1, R_1\} \text{ for } A_1, \ldots, \{P_k, R_k\} \text{ for } A_k.$$

Recognition

For each input pattern A, compute k values of the binding energy:

$$w_1 = -P_1^T A R_1, \ldots, w_k = -P_k^T A R_k.$$

Determine the class index by the minimal value of the binding energy according to (3.12):

$$c : w^* = \min_c \{w_c\}. \tag{3.12}$$

3.2.3 Unsupervised Learning

Consider a matrix $A = [X_1 \ldots X_m]^T$ of dimension $m \times n$ formed by the m vectors (patterns) $X_1, \ldots, X_m$. Consider the SVD of this matrix:

$$A = s_1 \begin{bmatrix} p_{11} \\ \cdots \\ p_{1n} \end{bmatrix} R_1^T + s_2 \begin{bmatrix} p_{21} \\ \cdots \\ p_{2n} \end{bmatrix} R_2^T + \cdots, \quad (3.13)$$

where s_1, s_2 are the first two singular values, and R_1, R_2 are right singular vectors.

Note that every string i of the matrix A represents the values x_{ij} of n characteristics of the pattern X_i, where $i=1,\ldots,m$ and $j=1,\ldots,n$. Hence, according to the SVD properties, the components p_{1i}, p_{2i} of the left singular vectors P_1, P_2 satisfy the following equations:

$$p_{1i} = P(X_i^T) \frac{|X_i|}{s_1} R_1,$$
$$p_{2i} = P(X_i^T) \frac{|X_i|}{s_2} R_2,$$
(3.14)

where

$$|X| = \sqrt{x_1^2 + \cdots + x_n^2},$$
$$P(X) = \frac{X}{|X|}.$$

Comparison of (3.4) and (3.14) makes it obvious that the components p_{1i}, p_{2i} can be computed as binding energies w_{1i}, w_{2i} between FP $P(X_i)$ and FPs R_1, R_2, respectively. Thus, every vector X_i with n characteristics is mapped to only two values of binding energies.

Such a mapping gives a mathematically rigorous way to represent and view all patterns, no matter how many indicators they have, as points in a two-dimensional space of binding energies $\{w_1, w_2\}$. This plane could be treated also as a *shape space* of the IC (DeBoer et al., 1992). Such a representation of patterns in the shape space of the IC allows them to be classified in a natural way by the groups (*clusters*) of the neighboring points. Such a classification can be performed by experts using supervised learning as well as by IC using unsupervised learning.

3.3 Main Computing Procedures

Typical features of the IC model of molecular recognition have two major differences from the known approaches to pattern recognition. First, within the proposed model, sets of indicators are not coded straight, but they only determine the binding energy between definite FPs (let us call them *FP-probes*). Second, the similarity between these sets of indicators is determined not by comparison of indicators, but by recognition between FP-probes. As a result, the recognizing class is determined by those FP-probes that have the minimal binding energy (the best recognition).

Apart from its theoretical importance to the basic biological mechanisms of molecular recognition, the application of such a mathematical model to pattern recognition is interesting for the following reasons:

The central procedure of the model is SVD, which determines FP-probes (singular vectors) and their binding energy (singular values). The SVD of an arbitrary matrix A has been considered in detail in Section 2.4.2.

Therefore, the IC mathematical model of pattern recognition is based on the properties of the SVD. Consequently, the main computational procedures of the model are as follows:

1. The definition of a matrix that determines the binding energy between FPs;
2. Computing the SVD of the matrix to obtain FP-probes and the values of binding energy between them;
3. Computation of vector norms, which determine the specificity of recognition.

The central procedure, computing the SVD of the matrix (2.9), is realized by the *deflation method* (see Section 2.4.2).

From the other side of pattern recognition, the IC
model has to include the following main procedures:

- *Supervised learning* (learning with a teacher or expert);
- *Unsupervised learning* (automated classification);
- *Recognition*.

From this viewpoint, computing the SVD represents the central procedure and determines the learning. Moreover, variants in the formation of the initial matrix and computation of vector norms give variants in learning and recognition. Let us consider those variants.

3.3.1 Supervised Learning

Let a *learning sample* $\{X_1, X_2,...\}_k$ be given with the same class $z=k$ assigned by the experts to every vector. Let such samples be provided for all classes $k=1,...,k_c$ (k_c is the general number of classes, defined by experts). The learning problem is solved in two stages. In the first stage, a matrix $A_{\{k\}}$ is formed, which determines the binding energy between the FPs. In the second stage, the SVD

(2.9) of each such matrix is computed. As a result, we determine FP-probes $(L_1, R_1, \ldots, L_p, R_p)_k$ and binding energies between them $(s_1, \ldots, s_p)_k$ for every class.

When forming matrices $A_{\{k\}}$, two tasks have to be solved. First, it is necessary to select the dimension of the matrices to which vectors of learning samples will be folded. Second, for every class it is necessary to form the best matrix, which takes into account all vectors of the learning sample for this class.

In accordance with Theorem 3.2, folding a vector to a matrix increases the specificity of recognition. Let n be the dimension of the vector of indicators: $X=[x_1 \ldots x_n]^T$. Then the dimension of the matrix, which determines the binding energy, can be $n_i \times n_j$, where n_i is the number of rows and n_j is the number of columns in the matrix. Hence, the condition $n_i n_j = n$ has to be satisfied, which means that the possible dimension of the matrix is determined by the complete set of divisors of the dimension of the vector of indicators. If we have several divisors, then the choice is determined by the possibility of obtaining the maximal number of decomposition elements (2.9), which is equal to the rank r of the matrix and is restricted by the condition $r \leq \min\{n_i, n_j\}$. Hence, our choice is determined by the conditions

$$\max(\min\{n_i, n_j\}),$$
$$n_i n_j = n.$$

Note that the rank of the matrix can be increased by an artificial mode, by introducing additional indicators (which have, for example, the same value for all vectors). Nevertheless, in any case, we can also restrict ourselves to the variant of vector representation (with the dimension of the matrix $n \times 1$ or $1 \times n$).

Note also that in principle, for different classes, matrices with different dimensions could be selected. However, this case needs additional procedures for recognition. Therefore, consider that matrices for all classes have the same dimension.

Now consider the task of forming the matrix A that has the best correspondence with all learning vectors of given class $X^{\{1\}}, \ldots, X^{(m)}$ (here and in the sequel we will omit the index of the class for convenience). Designate by $A^{\{1\}} = [a_{ij}^{(1)}], \ldots, A^{(m)} = [a_{ij}^{(m)}]$ matrices of corresponding dimension, which represent folded vectors of the learning sample.

Formulate the task as follows: Determine a value s and vectors $X=[x_i]^T$, $Y=[y_j]^T$, such that they give the best approximation (by least squares criteria) of all learning matrices

$$A^{(1)} \cong sXY^T, \ldots, A^{(m)} \cong sXY^T.$$

Strictly speaking, it is required to minimize a quadratic form,

$$Q = \sum_{i,j}\left(a_{ij}^{(1)} - sx_i y_j\right)^2 + \cdots + \sum_{i,j}\left(a_{ij}^{(m)} - sx_i y_j\right)^2.$$

The necessary condition of minimum Q is that its partial derivatives with respect to the variable s should equal zero: $Q'_X = 0, Q'_Y = 0, Q'_s = 0$. From these equalities the following system of vector–matrix equations is derived:

$$msX(Y^T Y) = [A^{(1)} + \cdots + A^{(m)}]Y,$$
$$ms(X^T X)Y^T = X^T[A^{(1)} + \cdots + A^{(m)}], \quad (3.15)$$
$$ms(X^T X)(Y^T Y) = X^T[A^{(1)} + \cdots + A^{(m)}]Y.$$

Suppose that the vectors are unit vectors, $X^T X=1$, $Y^T Y=1$, and define the matrix A as the mean of learning matrices:

$$A = \frac{1}{m}[A^{(1)} + \cdots + A^{(m)}]. \quad (3.16)$$

Then, obvious solutions of the system (3.15) are singular vectors and corresponding singular values of the matrix (3.16). Hence, the choice of matrix A as the mean of learning matrices for a given class is theoretically the best.

In this context, consider the task of *retraining*.

Let r singular values and vectors be determined by the optimal matrix (3.16). Let us adjust this matrix by the addition of expert information $A^{(m+1)}$ obtained for the given class. Then the task of optimal retraining is solved by the recursive computation of the arithmetic mean:

$$A = \frac{m}{m+1}(s_1 P_1 R_1^T + \cdots + s_r P_r R_r^T) + \frac{1}{m+1} A^{(m+1)}.$$

The final stage of learning by the SVD of the optimal matrix has been described in detail above. Note that the choice of the number p of the elements of decomposition cannot exceed the rank of the matrix: $p \leq r$. So, if $r \geq 3$, then $p=3$ is sufficient for all practical purposes.

After computing the SVD, there is no need to store the matrix itself, because it can be retrieved at any time by decomposition (2.9). Moreover, the SVD of the matrix gives at the same time an optimal filter of expert information, because the elements of decomposition (2.9) with small singular values correspond, in fact, to informational "noise."

3.3.2 Unsupervised Learning

Let a learning sample $X_1,\ldots,X_m$ be given, and assume that no class indication is given a priori for these vectors. Based on this sample, partitioning is required for the whole n-dimensional indicator space $\{X\}$ on classes $\{X\}_k$, where the number of classes k is generally unknown.

Reduce this task to the task that has been solved in Section 3.3.1. For this purpose reformulate the task of retraining as the task of partitioning the whole learning sample into classes $\{X_1, X_2, ...\}_k$, $k=1,...,k_c$.

Form the matrix $A = [X_1 ... X_m]^T$, whose rows are vectors of the learning sample. Obtain the SVD (2.9) of this matrix, restricted by p elements of decomposition, where $p \leq r$. As a result, for any n-dimensional vector X_j, $j=1,...,m$, of the learning sample we have a corresponding p-dimensional vector

$$W_j : X_j \to W_j,$$
$$W_j = [l_{1j} \, l_{2j} ... l_{pj}]^T,$$

where $l_{1j},...,l_{pj}$ are the jth coordinates of the left singular vectors:

$$L_1 = [l_{11} ... l_{1j} ... l_{1m}]^T, ..., L_p = [l_{p1} ... l_{pj} ... l_{pm}]^T.$$

The physical sense of the vector W_j is given by its coordinates. It consists of the values of the binding energy between FP-probes, corresponding to the right singular vectors $R_1,...,R_p$ and the FP-sample, corresponding to the vector X_j:

$$l_{1j} = w(R_1, X_j), ..., l_{pj} = w(R_p, X_j).$$

The geometric sense of this transformation consists in the representation of initial vectors $X_1, ..., X_m$ in the new coordinate system, whose axes $R_1,...,R_p$ are theoretically optimal (in the sense of minimum squares) with respect to the whole learning sample. Moreover, the dimension of this new representation is essentially smaller than the dimension of the initial vectors $p<n$. Therefore, the main practical significance of this representation is its ability to detect and display clearly the natural groups (classes) of initial vectors (if such groups exist) as points with coordinates $\{W\} = \{w_1, w_2, w_3\}$ within the unit cube. Such natural classes are self-detecting, mainly by using the SVD as an optimal compression of learning information, which removes "noise."

Formally, the aforementioned classes can be detected, based on the Euclidean norm, by a matrix of distances $D=[d_{ab}]$ between every pair of vectors:

$$d_{ab} = \text{dist}_E(W_a, W_b), \quad a,b=1,...,m.$$

For example, classes $\{W\}_k$ could be detected by proceeding from the condition of compactness, which means that the distance between any two vectors of the same class has to be lower than the distance between any two vectors of different classes:

$$\text{dist}_E(W, W_a) < \text{dist}_E(W, W_b), \forall \{W, W_a, W_b\}: W \in \{W\}_k, W_a \in \{W\}_k,$$
$$W_b \in \{W\}_i, k \neq i.$$

Theoretically, any such problem always has at least one solution:

$$\{W\}_k = W_k,$$
$$k = 1, \ldots, k_c,$$

where every learning vector forms a separate class.

If necessary, we can use other, more refined, methods of automatic partitioning of the new indicator space $\{W\}$, for example, the *dynamic cores method* (Faure, 1985). In any case, when partition $\{W\}_k$ has been obtained, it induces, obviously, the corresponding partition of the initial vectors $\{X\}_k$, where $X \to W$, $W \in \{W\}_k$ induces $X \in \{X\}_k$. Therefore, the unsupervised learning is reduced to the problem of supervised learning, which has already been solved.

It is worthwhile mentioning that in practice it is reasonable to use not fully automated, but combined learning. In this case initial learning samples $X_1, \ldots, X_m$ are transformed to the indicator space $\{W\}$ and then presented to experts (for example, on a computer display). This clear representation makes possible effective detecting of separate groups of points that correspond to different classes.

3.3.3 Recognition

Assume that the learning problem has been solved and the indicator space $\{X\}$ has been partitioned into the classes (indices) $z_k = \{X\}_k$, $k = 1, \ldots, k_c$, where the FP-probes $(L_1, R_1, L_2, R_2, \ldots, L_p, R_p)_k$ and the standard binding energies $(s_1, s_2, \ldots, s_p)_k$ are determined for every class k. Then recognition consists in assigning an index $z = z(\widetilde{X})$, $z \in \{z_k\}$, to an arbitrary vector of indicators $\widetilde{X}$. Consider a decision rule for this problem.

Fold the initial vector $\widetilde{X}$ to a matrix $\widetilde{A} = \widetilde{A}(\widetilde{X})$ of dimension $n_i \times n_j$, where n_i, n_j are dimensions of the left and the right singular vectors L and R, respectively.

Define the distance $d_k(\widetilde{X})$ between the class k and the initial vector $\widetilde{X}$ by the following formula:

$$d_k(\widetilde{X}) = \sum_{i=1}^{p} \left\| \widetilde{A}(R_i)_k - (s_i)_k (L_i)_k \right\|_E + \sum_{i=1}^{p} \left\| \widetilde{A}^T (L_i)_k - (s_i)_k (R_i)_k \right\|_E. \quad (3.17)$$

Note that $d_k(\widetilde{X}) \geq 0$ for any vector $\widetilde{X}$ and any class $k = 1, \ldots, k_c$. Specifically, $d_k(\widetilde{X}) = 0$ if and only if $\widetilde{A}$ is the learning matrix for the class k.

Based on this function of distance (3.17), the decision rule assigns a class to the vector $\widetilde{X}$ that has the minimal distance to this vector among all classes

$$z(\widetilde{X}) = z_k : \min_k \{ d_k(\widetilde{X}) \}.$$

Note that the distance function (3.17) is not the only possible variant. For example, we can use only one of the sums in the right side of (3.17) as a distance. Using two sums of the right side (3.17) is reasonable, because thus we involve all the information contained in the SVD (2.9). In addition, the Euclidean norm is not compulsory, because for some cases the Manhattan norm turns out to be more appropriate. In any case, it is worth emphasizing that the choice of the norm for the decision rule is mainly determined not by theoretical properties of the norm, but by practical features of the application. Speaking concretely, this choice is determined by the appearance of the groups of points in the new indicator space $\{W\}$.

Another more precise definition concerns the designation of classes. Theoretically, it is sufficient for recognition to designate classes z_k only by their indices $k=1,2,...,k_c$. However, together with numeric indices, it is convenient to use symbolic notation for the classes. For example, for a task of complex ecological evaluation it could be $z_1=$ '*good*', $z_2=$ '*bad*', while for a task of information security it could be $z_1=$ '*self*', $z_2=$ '*nonself*'.

4
Language Representation and Knowledge Based Reasoning

From the algebraic point of view, any language represents some subset of words of a free monoid (Lalleman, 1979). Formally, the main problems of language representation and knowledge-based reasoning can be reduced to the limitations on a free monoid, and to those that also allow such words and sentences to be recognized (Salomaa, 1981). By their nature, at least, these problems are similar to those of representation and recognition of self and nonself by the natural immune system. This chapter will show that the analogy is actually much more profound, and that it leads to developing an IC approach to language representation and knowledge-based reasoning.

4.1 Morphology: Peptide Spectrum of a Word

Consider an alphabet $A1 = \{a,b,...\}$. Let two fixed torsion angles and a unitary quaternion $Q(\varphi, \psi)$ correspond to each letter of the alphabet, according to (2.1), (2.2):

$$a \to Q_a(\varphi_a, \psi_a),\ b \to Q_b(\varphi_b, \psi_b),\ ... .$$

Let an FP correspond to each word in the alphabet. Let the Q-vector of such an FP be determined by a product of quaternions according to the sequence of letters of the given word, for example,

$$abcd \to Q_{abcd} = Q_a Q_b Q_c Q_d.$$

Thus, some value of the free energy (2.3) of an FP corresponds to each word of the alphabet.

Example 4.1 Assume an alphabet of $n \geq 2$ characters. Let the following quaternion correspond to each letter of the alphabet:

$$Q_x(\varphi) = \left(\cos\frac{\varphi}{2},\ \sin\frac{\varphi}{2}\right), \varphi : \cos\varphi = \frac{1}{n}.$$

Let the energy matrix $V^{(i)}$ correspond to the ith link of the FP:

$$V^{(1)} = \begin{bmatrix} -1 & 0 \\ 0 & 1 \end{bmatrix},$$

$$V^{(i)} = \begin{bmatrix} -v^{(i-1)} & 0 \\ 0 & v^{(i-1)} \end{bmatrix},$$

where $v^{(i-1)}$ is the free energy of the previous link, $i>1$. Then for any word of length m, the free energy is obtained as

$$v = v^{(m)} = \frac{1}{n^m}.$$

The obtained value is known as the *code indicator of the word* (Salomaa, 1981). Thus the code indicator for a language, L, designated as ci(L), is defined as the sum of code indicators of all words of the language, and if the language is code, then ci(L)≤1.

Example 4.1 demonstrates that the code indicator can be represented as the free energy of an FP.

Definition 4.1 An *admissible word* is a word for which the free energy of the corresponding FP does not exceed a particular threshold.

Example 4.2 Let each letter of an alphabet correspond to a monopeptide with nonzero controls $v_{11}=v_{22}=v_{33}=v_{44}=-1$, and all remaining controls equal to zero. Then each letter of the alphabet has a corresponding free energy $v=1$.

Let each word with a number of characters $n>1$ correspond to a complex of n monopeptides with binding energies equal to zero. Then, according to (2.7), the energy of such a complex is equal to $v=n$. Simply put, the energy of the complex is equal to the number of letters in any word. Consider a given threshold n_h; then any word of length $n>n_h$ is inadmissible.

Example 4.2 gives a possible mechanism for the limitation of word length in formal systems. This is important because in usual formal grammars (Ginsburg, 1966) or systems of formal logic (Thayse et al., 1988) there are no limitations on the number of sequentially used axioms or rules. This causes the appearance of unlimited words or sentences, which are nevertheless "correctly formed." This circumstance is marked as a serious inconsistency (Thom, 1975), since in

4.1 Morphology: Peptide Spectrum of a Word

natural languages the number of letters in a word and of words in a sentence is always limited.

Let us also show how admissible combinations of letters (*morphology*) can be set through the use of FPs.

Theorem 4.1 In an alphabet of 2 characters, for any subset of words of length 2 there exists a dipeptide and a threshold that make admissible this and only this subset of words.

We omit the proof of the theorem for economy of space (see Tarakanov, 1999a).

Example 4.3 Consider an alphabet of four characters, which are round and square brackets: (,) , [,]. Let the following quaternions correspond to the brackets:

$$(\to Q\left(0, \frac{\pi}{2}\right),$$

$$) \to Q\left(-\frac{\pi}{2}, -\pi\right),$$

$$[\to Q\left(-\frac{\pi}{2}, -\frac{\pi}{2}\right),$$

$$] \to Q\left(\frac{\pi}{2}, -\frac{\pi}{2}\right).$$

Assume that the only nonzero controls of a dipeptide are $v_{22}=v_{33}=1$. Then the energies of the dipeptide for all possible words of length 2 over the given alphabet are presented in Table 4.1.

Now let the threshold equal $v_h=-1$. Then the only admissible words are correctly set brackets: ( ) and [ ], for which the energy is equal to $v=-1$.

Now consider a possible mechanism for the definition of admissible words of arbitrary length.

Definition 4.2 Consider word y as a subword of some word xyz, where probably $x=\varnothing$ or $z=\varnothing$, but $xz \neq \varnothing$, where $\varnothing$ designates an empty word. Define *deletion* as the following conversion of the word xyz:

$$\text{del}(y): xyz \to xz.$$

word →	energy	word →	energy	word →	energy	word →	energy
((	$-\frac{3}{4}$	)(	$-\frac{1}{4}$	[(	$-\frac{1}{8}$	](	$-\frac{3}{8}$
()	-1	))	$-\frac{3}{4}$	[)	$-\frac{1}{2}$	])	$-\frac{1}{2}$
([	$-\frac{1}{2}$	)[	$-\frac{3}{8}$	[[	$-\frac{3}{4}$	][	$-\frac{1}{4}$
(]	$-\frac{1}{2}$	)]	$-\frac{3}{8}$	[]	-1	]]	0

Table 4.1 Energy of a dipeptide for all two-letter words in the four-bracket alphabet.

Definition 4.3 The *processing* of a word is a sequence of all possible deletions of its admissible subwords.

Let, according to Example 4.3, the admissible two-letter words in the four-bracket alphabet be only the correctly set brackets () and []. Consider the processing of some word of arbitrary length in this alphabet, for example,

$$)[([])() \to)[() \to)[.$$

The above considerations are constructive in determining whether the word of arbitrary length that arose from this processing is admissible or inadmissible. In this case the admissible words of arbitrary length in the four-bracket alphabet are equivalent to the two-letter Dyck language D_2, which is known as a common generator of context-free languages (Ginsburg, 1966). Thus the processing is in fact an analogue of the Dyck map, or erasing morphism (Salomaa, 1981).

Two important obstacles are worth noting. First, processing in nature is realized at a biomolecular level (Alberts et al., 1986). Second, processing systems give more economic and informative descriptions of structural classes than generative grammars (Goldfarb, 1992).

Definition 4.4 The *peptide spectrum* of a word is the $m \geq 1$ FPs that correspond to the word.

If a peptide spectrum is determined for each word of the given alphabet, then the vector $F=[v_i]$ corresponds to each word, where v_i is the free energy of the ith FP, $i=1,..., m$. In particular, a peptide spectrum corresponds to a free semigroup mapped to the Euclidean space of dimension m, and thus allows the possibility of interpreting operations over words in terms of linear algebra operations.

4.2 Syntax: Matrix Eigenlanguages

Based on the considerations of the previous section, any chain of linguistic symbols can be considered as an FP. Thus it is natural to treat the admissible subwords as subpeptides (or *secondary structures*) of an FP. In this connection, consider cases in which any FP corresponding to a word of some language contains strictly defined combinations of the secondary structures. As will be shown later, these cases lead to a generalization of the concepts of the eigenvalues and eigenvectors of the matrices over linguistic symbols.

Let us introduce the necessary notation.

Designate a *semiring* as $\Omega = \langle Al, +, \bullet \rangle$, where Al is an alphabet with two operations, $+$ is an association of words, and $\bullet$ is a catenation of words (the symbol for this operation is usually omitted).

Designate special elements of the semiring as $\varnothing$ (zero), ε (unit), where $\varnothing \neq \varepsilon$. These elements satisfy the following axioms (Lalleman, 1979):

1. $\langle Al, +, \varnothing \rangle$ is a commutative monoid with a neutral element $\varnothing$;
2. $\langle Al, \bullet, \varepsilon \rangle$ is a monoid with a neutral element ε;
3. $x(y+z) = xy + xz$, $(y+z)x = yx + zx$, $\forall x,y,z \in \Omega$;
4. $x\varnothing = \varnothing x = \varnothing$, $\forall x \in \Omega$.

Designate by Ω_n a set of vectors of dimension n, and by $\Omega_{n,n}$ the set of matrices of dimension $n \times n$ over the semiring Ω, where

$$\Omega_n = \Omega \times \Omega \times \cdots \times \Omega \ (n \text{ times}),$$
$$\Omega_{n,n} = \Omega_n \times \Omega_n \times \cdots \times \Omega_n \ (n \text{ times}).$$

Designate by $A=[a_{ij}]$, $A \in \Omega_{n,n}$, a square matrix in which elements $a_{ij} \in \Omega$, $i,j=1,...,n$, are associations of words (languages) of the semiring.

Represent a language L in vector–matrix form:

$$L = Y^T A X, \qquad (4.1)$$

where $Y, X \in \Omega_n$ are vectors of dimension n over the semiring.

Example 4.4 Consider an automaton grammar in vector–matrix notation over a semiring of the formal power series (Kuich and Salomaa, 1986):

$$X = AX + \hat{X},$$

where X, $\hat{X}$ are vectors of *nonterminals* and *terminals* of the grammar, and A is a matrix of transitions (between states of a finite automaton) whose elements are letters of the alphabet. Then the solution of this equation is determined by the expression

$$X = A^* \hat{X}$$
$$A^* = [E - A]^{-1},$$

where E is an analogue of the unit matrix with diagonal elements ε and all nondiagonal elements equal to $\varnothing$.

Let the first nonterminal x_1 be an axiom (initial symbol) of the grammar. Then it is possible to determine the language generated by such a grammar in the form (4.1) as follows:

$$x_1 = [\varepsilon \varnothing \dots \varnothing] A^* \hat{X}.$$

In fact, Example 4.5 proves that languages of the form (4.1) exist.

Now let matrix A and vectors X, Y in representation (4.1) satisfy at least one of the following equations:

$$AX = \lambda_X X, \qquad (4.2)$$
$$Y^T A = Y^T \lambda_Y, \qquad (4.3)$$
$$AX = X \lambda_S, \qquad (4.4)$$
$$Y^T A = \lambda_P Y^T, \qquad (4.5)$$

where $\lambda_X, \lambda_Y, \lambda_S, \lambda_P \in \Omega$. Then substitutions (4.2) through (4.5) interpret the language (4.1) as one of the following linear algebra forms:

$$L = \{ Y^T \lambda X \cup Y^T X \lambda \cup \lambda Y^T X \}, \quad \lambda \in \Omega;$$

i.e., any word of the language necessarily includes a combination of two subwords $\omega_Y \in y_i$, $\omega_X \in x_i$, and any combination of the subwords $\omega_Y \in y_i$, $\omega_X \in x_j$ for $j \neq i$ is inadmissible.

Definition 4.5 If λ_X and nonzero vector X satisfy (4.2), then call λ_X a *right eigenelement* and X a *right eigenvector* corresponding to λ_X, and call $\langle \lambda_X, X \rangle$ a *right eigenlanguage* of the matrix A. Accordingly, call the solutions of (4.3) through (4.5) *left* (4.3), *suffix* (4.4), and *prefix* (4.5) eigenlanguages.

The following two examples show that such eigenlanguages exist, but not for every matrix.

Example 4.5 Consider the matrix

$$A = \begin{bmatrix} a_1 & a_2 \\ \varnothing & a_1 \end{bmatrix},$$

where $a_1, a_2 \in \Omega$. Then this matrix has the following eigenlanguages:

$$\lambda_X = \lambda_Y = \lambda_S = \lambda_P = a_1,$$
$$X = [x_1 \; \varnothing]^T,$$
$$Y = [\varnothing \; y_2]^T,$$
$$S = [(a_1)^* \; \varnothing]^T,$$
$$P = [\varnothing \; (a_1)^*]^T.$$

Example 4.6 Consider the matrix

$$A = \begin{bmatrix} a & b \\ c & a \end{bmatrix},$$

where a,b,c are characters (one-letter words). Then (4.1) looks like

$$ax_1 + bx_2 = \lambda_X x_1,$$
$$cx_1 + ax_2 = \lambda_X x_2.$$

From the first equation of this system one can deduce $\lambda_X = a + bz_1$, $z_1 \in \Omega$, and $x_2 = z_1 x_1$. From the second equation one can deduce $\lambda_X = a + cz_2$, $z_2 \in \Omega$, $x_1 = z_2 x_2$. Hence $x_1 = z_2 x_2 = z_2 z_1 x_1$ and $z_1 = z_2 = \varepsilon$. However,

$$a + bz_1 \neq a + cz_2.$$

The last equation means that the given matrix has no right eigenlanguage.

Examples 4.5, 4.6 allow us to determine the conditions under which a square matrix over a semiring has an eigenlanguage and to develop methods of solving (4.2) through (4.5) to determine this eigenlanguage.

Definition 4.6 A matrix is said to be *right (left) degenerate* if all elements of at least one column (row) of the matrix are equal to zero.

Proposition 4.1 A matrix over a semiring has zero eigenelements if and only if the matrix is right (left) degenerate. Thus if the *j*th column or row of the matrix is zero, then the matrix has an eigenvector whose *j*th component is an arbitrary nonzero language, and all remaining components are equal to zero.

We omit the proof of Proposition 4.1 for economy of space (see Tarakanov, 1999a).

Let $I=(i_1,...,i_k)$ be some subset of natural numbers; set $N = (1, ..., n)$, and $i_1 < i_2 < \cdots < i_k$, $k<n$, where $n \times n$ is the dimension of a matrix A. Introduce the following languages over the set of elements of the matrix A:

$$L_{XS}(i_1,...,i_k) = \sum_{\substack{i=i_1,...,i=i_k \\ j \neq i_1,...,j \neq i_k}} a_{ij},$$

$$L_{YP}(i_1,...,i_k) = \sum_{\substack{i \neq i_1,...,i \neq i_k \\ j=i_1,...,j=i_k}} a_{ij}.$$

While the appearance of these languages is not of any importance, the main challenge is in determining whether they are equal to zero. This is achieved by verifying the following properties of the matrix corresponding to $L_{XS}(i_1,...,i_k)=\varnothing$:

1. Delete columns $i_1,...,i_k$;
2. Delete all strings except strings $i_1,...,i_k$;
3. All remaining elements of the matrix should be zero.

Similarly, the condition $L_{YP}(i_1,...,i_k)=\varnothing$ gives the same property, but for the transpose matrix A^T.

Definition 4.7 A matrix is said to be *right (left) complete* if for any subset of strings (columns) $I \in N$, we have $L_{XS}(I)=\varnothing$ ($L_{YP}(I)=\varnothing$).

The following statements can be derived immediately from Definition 4.7:

1. A degenerate matrix cannot be complete;
2. A complete matrix cannot be degenerate.

Note also that an incomplete matrix can be nondegenerate (see Example 4.5), and a complete matrix can have zero components (see Example 4.7 below for $a_{12} \neq \varnothing$, $a_{32} \neq \varnothing$, $a_{13} \neq \varnothing$).

The following statement gives the main property of a complete matrix.

Proposition 4.2 For a right (left) complete matrix there is no right or suffix (left or prefix) eigenvector with any zero component.

$$A^T = Y_1 \omega_1^{-1} X_1^T + \cdots + Y_n \omega_n^{-1} X_n^T, \qquad (4.6)$$

where $X_i, Y_i \in \Omega_n$ are some vectors of dimension n over a semiring. It is obvious that for any matrix A such a decomposition exists; for example, for $\omega_i = \varepsilon$, $Y_i = A_i^T$, $X_i = E_i$, or for $\omega_i = \varepsilon$, $Y_i = E_i$, $X_i = A_i$, where A_i^T and A_i are vectors whose components are, respectively, the ith string and the ith column of the matrix A, and E_i is the ith column of the identity matrix.

The decomposition (4.6) involves the vectors

$$X = X_1 + \cdots + X_n,$$
$$Y = Y_1 + \cdots + Y_n.$$

Consider also the following conditions:

$$\{\forall i = 1,\ldots,n: \; Y_i = A_i^T, \; X_i = \omega_i E_i, \; Y_i^T X \omega_i^{-1} = \lambda_X, \; \omega_i, \lambda_X \in \Omega: \}, \qquad (4.7)$$

$$\{\forall i = 1,\ldots,n: \; X_i = A_i, \; Y_i = \omega_i E_i, \; \omega_i^{-1} Y^T X_i = \lambda_Y, \; \omega_i, \lambda_Y \in \Omega: \}. \qquad (4.8)$$

Definition 4.8 Call the decomposition (4.6) *right* if the conditions (4.7) are satisfied, or *left* if the conditions (4.8) are satisfied.

The following example proves the existence of right and left decompositions for some complete matrices.

Example 4.9 Consider two matrixes

$$A = \begin{bmatrix} a & b \\ bc^2 & a \end{bmatrix},$$

$$B = \begin{bmatrix} a & ca \\ b & cb \end{bmatrix}.$$

For the first matrix there exists a right decomposition, and for the second matrix there exists a left decomposition:

$$A^T = \begin{bmatrix} a \\ b \end{bmatrix} [\varepsilon \; \varnothing] + \begin{bmatrix} bc^2 \\ a \end{bmatrix} c^{-1} [\varnothing \; c],$$

$$B^T = \begin{bmatrix} \varepsilon \\ \varnothing \end{bmatrix} [a \; b] + \begin{bmatrix} \varnothing \\ c \end{bmatrix} c^{-1} [ca \; cb].$$

4.2 Syntax: Matrix Eigenlanguages

We omit the proof of Proposition 4.2 for economy of space (see Tarakanov, 1999a).

However, incomplete matrices can have eigenvectors with zero components, as shown in the following example.

Example 4.7 Consider a matrix with all nonzero elements, except $a_{12}=a_{13}=\varnothing$:

$$A = \begin{bmatrix} a_{11} & \varnothing & \varnothing \\ a_{21} & a_{22} & a_{23} \\ a_{31} & a_{32} & a_{33} \end{bmatrix}.$$

For this matrix $L_{XS}(1)=\varnothing$, and it can have an eigenvector $X=[\varnothing \ x_2 \ x_3]^T$, whose nonzero components $x_2 \neq \varnothing$, $x_3 \neq \varnothing$ are determined only by a complete submatrix

$$A = \begin{bmatrix} a_{22} & a_{23} \\ a_{32} & a_{33} \end{bmatrix}.$$

Suppose now that $a_{32}=\varnothing$. Then

$$A = \begin{bmatrix} a_{11} & \varnothing & \varnothing \\ a_{21} & a_{22} & a_{23} \\ a_{31} & \varnothing & a_{33} \end{bmatrix},$$

$$L_{XS}(1,3) = \varnothing,$$

and this incomplete matrix has an eigenlanguage

$$\langle \lambda_X = a_{22}, X=[\varnothing \ x_2 \ \varnothing]^T \rangle,$$

where x_2 is an arbitrary nonzero language. Thus a complete submatrix of such a matrix that defines nonzero component x_2 is the element a_{22}. Note that inequality $L_{XS}(3) \neq \varnothing$ is satisfied independently from equalities $L_{XS}(1,3)=\varnothing$, $L_{XS}(1)=\varnothing$, because $a_{33} \neq \varnothing$.

These results allow us to formulate an algorithm for determining eigenlanguages (see Tarakanov, 1999a).

Now consider the problem of determining the eigenlanguages of a complete matrix.

Expand a semiring Ω to a field $\Xi=\Xi(\Omega)$ by putting in correspondence to each element $\omega \in \Omega$ an inverse element $\omega^{-1} \in \Xi \setminus \Omega$ such that $\omega^{-1}\omega = \omega\omega^{-1} = \varepsilon$. For an arbitrary matrix $A \in \Omega_{n,n}$ introduce the decomposition

Thus for the first matrix

$$Y_1^T X \omega_1^{-1} = Y_2^T X \omega_2^{-1} = a + bc, \ Y_1 = [a\ b]^T, \ Y_2 = [bc^2\ a]^T, X = [\varepsilon\ c]^T,$$

for the second matrix

$$\omega_1^{-1} Y^T X_1 = \omega_2^{-1} Y^T X_2 = a + cb, \ X_1 = [a\ b]^T, X_2 = [ca\ cb]^T, Y = [\varepsilon\ c]^T,$$

and for both matrices $\omega_1 = \varepsilon$, $\omega_2 = c$.

The following proposition connects right and left decompositions with eigenlanguages of complete matrices.

Proposition 4.3 A complete matrix has a right (left) eigenlanguage if and only if a right (left) decomposition exists.

The proof of Proposition 4.3 can be found in (Tarakanov, 1999a).

As an illustration of Proposition 4.3, note that the matrices considered in Example 4.8, have the following eigenlanguages:

$$A: \langle \lambda_X = a+bc, X=[\varepsilon\ c]^T \rangle,$$
$$B: \langle \lambda_Y = a+cb, Y=[\varepsilon\ c]^T \rangle.$$

The following obvious properties hold for right and left eigenlanguages:

$$\forall \omega \in \Xi \ (Z_X = X\omega, AX = \lambda_X X) \Rightarrow AZ_X = \lambda_X Z_X,$$
$$(Z_Y = \omega Y, Y^T A = Y^T \lambda_Y) \Rightarrow Z_Y^T A = Z_Y^T \lambda_Y.$$

These properties mean that a right (left) eigenvector remains an eigenvector after adding to all its components from the right (left) side any element of a field Ξ. Therefore, the eigenvector can be "normalized" so that any ith component will be a unit that can be obtained by an operation $Z_X = Xx_i^{-1}$ or $Z_Y = y_i^{-1} Y$. However, some other components of the normalized eigenvector can appear as elements of a set $\Xi \backslash \Omega$; i.e., they can be from a semiring.

Decomposition (4.6) and Proposition (4.3) allow us to reduce the determination of right and left eigenlanguages of the given matrix to the search for appropriate elements $\omega_1, ..., \omega_n$, while the property of normalization determines the direction of this search as follows.

Let $\omega_1 = \varepsilon$. Then according to condition (4.7) a right eigenlanguage should satisfy the condition $a_{i1} \omega_i^{-1} \in \Omega$; i.e., an element $\omega_i \in \Omega$ should be a termination (proper or improper) for an element a_{i1} of a matrix A. Similarly, according to condition (4.8), a left eigenlanguage should have an element ω_i that can be selected only from the initial sublanguages of an element a_{1i}; i.e., $\omega_i^{-1} a_{1i} \in \Omega$.

Now let $\omega_i \in \exists \backslash \Omega$. Then condition (4.7) restricts the element ω_l so that $a_{1i} \omega_i \in \Omega$; i.e., an inverse element $\omega_i^{-1} \in \Omega$ should be a termination of the element a_{1i}. Similarly, condition (4.8) in the case $\omega_i^{-1} \in \Omega$ implies a restriction $\omega_i a_{i1} \in \Omega$.

Thus, possible values ω_i are selected from sets Ω_i^R, Ω_i^L for right and left eigenlanguages respectively, where

$$\Omega_i^R = \{\omega_i \in \Omega : a_{i1} \omega_i^{-1} \in \Omega\} \cup \{\omega_i^{-1} \in \Omega : a_{1i} \omega_i \in \Omega\}, \quad (4.9)$$
$$\Omega_i^L = \{\omega_i \in \Omega : \omega_i^{-1} a_{1i} \in \Omega\} \cup \{\omega_i^{-1} \in \Omega : \omega_i a_{i1} \in \Omega\}. \quad (4.10)$$

These conditions give an algorithm for determining right and left eigenlanguages (see Tarakanov, 1999a).

Let us illustrate this algorithm by the following example.

Example 4.9 Let us determine the eigenlanguages of the matrix from Example 4.8. For this purpose consider an initial decomposition

$$A^T = \begin{bmatrix} a & bc^2 \\ b & a \end{bmatrix} = \begin{bmatrix} a \\ b \end{bmatrix} [\varepsilon \quad \varnothing] + \begin{bmatrix} bc^2 \\ a \end{bmatrix} [\varnothing \quad \varepsilon].$$

This decomposition is not a right decomposition, since

$$[a \; b] X \neq [bc^2 \; a] X,$$
$$X = [\varepsilon \; \varepsilon]^T.$$

The search for eigenlanguages of a matrix of dimension 2×2 is carried out through the selection of only one element ω_2. For the initial matrix, possible values $\omega_2 \in \Omega$ are determined by an element $a_{12} = b$, while values $\omega_2^{-1} \in \Omega$ are determined by an element $a_{21} = bc^2$. Thus

$$\Omega_2^L = \{b, b^{-1}, (bc)^{-1}, (bc^2)^{-1}\},$$

but no element from this set satisfies (4.10). Hence, the initial matrix has no left eigenlanguage.

Now address the issue of prefix and suffix eigenlanguages of a complete matrix.

Consider the following decomposition for an arbitrary complete matrix A over a semiring:

$$A = S_1 P_1^T + \cdots + S_m P_m^T, \qquad (4.11)$$

where $A \in \Omega_{n,n}$, $S_i, P_i \in \Omega_n$. Obviously, for any matrix A this decomposition exists, for example, when $m=n$, $S_i=A_i$, $P_i=E_i$.

Introduce vectors $S = S_1 + \cdots + S_m$, $P = P_1 + \cdots + P_m$, and consider the following conditions:

$$P^T S_1 = \cdots = P^T S_m = \lambda_P, \qquad (4.12)$$

$$P_1^T S = \cdots = P_m^T S = \lambda_S. \qquad (4.13)$$

Definition 4.9 Decomposition (4.11) is a *prefix* if conditions (4.12) are satisfied or a *suffix* if conditions (4.13) are satisfied.

The existence of the prefix and suffix decompositions for some complete matrices is obvious from the following example:

Example 4.10 Consider arbitrary vectors $S, P \in \Omega_n$ over a semiring and generate a matrix $A = SP^T$. Thus, $P^T S = \lambda$, $\lambda \in \Omega$, and conditions (4.11) through (4.13) are satisfied. Hence, the representation of the matrix A as a product of vectors S and P^T is simultaneously its prefix and suffix decomposition. Then the prefix and suffix eigenlanguages of the matrix are $\langle \lambda, P \rangle, \langle \lambda, S \rangle$, because

$$P^T A = P^T S P^T = \lambda P$$
$$AS = SP^T S = S\lambda.$$

In general, the connection between the eigenlanguages of a matrix and its prefix or suffix decomposition gives the following proposition:

Proposition 4.4 A complete matrix has a prefix (suffix) eigenlanguage if and only if there exists a prefix (suffix) decomposition of the matrix.

We omit the proof of Proposition 4.4 for economy of space (see Tarakanov, 1999a).

It is important in determining prefix or suffix eigenlanguages that unlike right (left) decomposition (4.6), where the number of items n is always equal to the number of strings (columns) of a matrix, the number of items m in a prefix or suffix decomposition (4.11) can vary within the limits $1 \leq m \leq n \times n$. Example 4.10 shows that it can be $m=1$ for a matrix of arbitrary dimension. In the following example the cases $m=n$, $m>n$ are considered.

Example 4.11 Consider a matrix

$$A = \begin{bmatrix} (c+d)a & c(a+b) \\ (c+d)b & d(a+b) \end{bmatrix}.$$

The prefix decomposition of this matrix for $m=n=2$ is

$$A = \begin{bmatrix} (c+d)a \\ (c+d)b \end{bmatrix} [\varepsilon \quad \varnothing] + \begin{bmatrix} c(a+b) \\ d(a+b) \end{bmatrix} [\varnothing \quad \varepsilon],$$

where

$$P = [a+b \quad a+b]^T,$$
$$\lambda_P = P^T S_1 = P^T S_2 = P^T S_3 = (a+b)(c+d).$$

The prefix decomposition of the same matrix for $m=3$, i.e., for $m>n$, is

$$A = \begin{bmatrix} c+d \\ \varnothing \end{bmatrix} [a \quad \varnothing] + \begin{bmatrix} \varnothing \\ c+d \end{bmatrix} [b \quad \varnothing] + \begin{bmatrix} c \\ d \end{bmatrix} [\varnothing \quad a+b],$$

where

$$P = [a+b \quad a+b]^T,$$
$$\lambda_P = P^T S_1 = P^T S_2 = P^T S_3 = (a+b)(c+d).$$

Generalizing Examples 4.11 and 4.12, it is necessary to note that any decomposition (4.11) for $m<n$ can be reduced to a variant $m=n$, for example, by decomposition of some items,

$$S_j P_j^T = S_j P_{j1}^T + S_j P_{j2}^T,$$
$$P_{j1}^T P_{j2} = \varnothing.$$

And the opposite is also true: If (4.11) is satisfied for some j_1, j_2, i.e.,

$$S_{j1} = S_{j2}, \quad P_{j1}^T P_{j2} = \varnothing,$$

then the number of terms of the decomposition is reduced by an operation

$$S_{j1} P_{j1}^T + S_{j2} P_{j2}^T = (S_{j1} + S_{j2})(P_{j1} + P_{j2})^T.$$

However, decomposition (4.11) in general cannot be reduced to the variant $m=n$, as shown in Example 4.11.

Now consider an arbitrary pair of vectors S_j, P_j, $1 \leq j \leq m$, from decomposition (4.11). Replace this pair by vectors $S_j^\omega = S_j \omega^{-1}$, $P_j^\omega = \omega_j P_j$. Obviously, this replacement does not contradict (4.11), because $S_j^\omega \left(P_j^\omega\right)^T = S_j P_j$. According to (4.12), (4.13), introduce the following sets of admissible replacements for determining prefix and suffix decompositions:

$$\Omega_j^P = \{\omega_j : (P^\omega)^T S_j \omega^{-1} \in \Omega\}, \quad (4.14)$$

$$\Omega_j^S = \{\omega_j : \omega_j P_j^T S^\omega \in \Omega\}, \quad (4.15)$$

where

$$P^\omega = P_1 + \cdots + P_{j-1} + P_j^\omega + P_{j+1} + \cdots + P_m, \quad S^\omega = S_1 + \cdots + S_{j-1} + S_j^\omega + S_{j+1} + \cdots + S_m.$$

Condition (4.14) means the following: Let $\omega_j \in \Omega$. Then either ω_j is a termination of $\forall [S_j]_i$, $i=1,\ldots,n$, where $[S_j]_i$ is the ith component of the vector S_j, or if $\exists i: [S_j]_i \omega_j \in \Xi/\Omega$, then

$$\omega_j [P]_i [S_j]_i \omega_j^{-1} \in \Omega,$$

whence $\omega_j [S_j]_i^{-1} \in \Omega$ is a termination for $\omega_j [P]_i$.

Let $\omega_j \in \Xi/\Omega$. Then either $\omega_j^{-1} \in \Omega$ is a beginning of $\forall [P_j]_i$, or if $\exists i: \omega_j [P_j]_i \in \Xi/\Omega$, then $(\omega_j [P_j]_i)^{-1} \in \Omega$ is a beginning of $[S_j]_i \omega_j^{-1} \in \Omega$.

Condition (4.15) can be interpreted similarly.

These results allow us to formulate an algorithm for determining prefix and suffix eigenlanguages (see Tarakanov, 1999a).

Example 4.12 Consider a matrix

$$A = \begin{bmatrix} ab & cbab \\ cbab & ab \end{bmatrix}.$$

Determine its prefix eigenlanguages for $m=n$. The initial decomposition of this matrix is

$$A = \begin{bmatrix} ab \\ cbab \end{bmatrix} [\varepsilon \quad \varnothing] + \begin{bmatrix} cbab \\ ab \end{bmatrix} [\varnothing \quad \varepsilon].$$

76 Language Representation

Hence,

$$\Omega_1^P = \Omega_2^P = \{\, b,\, ab,\, bab \,\}.$$

The following prefix eigenlanguages are obtained by checking that the conditions (4.12) for the initial decomposition and for various combinations of $\omega_1 \in \Omega_1^P$, $\omega_2 \in \Omega_2^P$, are satisfied:

$$\langle \lambda = ab+cbab, P=[\varepsilon\ \varepsilon]^T \rangle,$$
$$\langle \lambda = ba+bcba, P=[b\ b]^T \rangle,$$
$$\langle \lambda = ab+abcb, P=[ab\ ab]^T \rangle,$$
$$\langle \lambda = ba+babc, P=[bab\ bab]^T \rangle.$$

For the last eigenlanguage,

$$S_1 = [b^{-1}\ c]^T,$$
$$S_2 = [\, c\ b^{-1}]^T,$$

where $S_1, S_2 \notin \Omega_2$, but $S_1, S_2 \in \Xi_2$.

Note that if a matrix, independent from its completeness or degeneracy, has a prefix (suffix) eigenlanguage, then the matrix has enumerable sets of prefix (suffix) eigenlanguages. From (4.4), (4.5) it follows that

$$P^T A^k = \lambda_P\, P^T A^{k-1} = \cdots = (\lambda_P)^k P^T,$$
$$A^k S = A^{k-1} S \lambda_S = \cdots = S(\lambda_S)^k,\ \forall k = 1, 2, \ldots.$$

Thus, if a matrix A has a prefix or suffix eigenlanguage $\langle \lambda_P, P \rangle$ or $\langle \lambda_S, S \rangle$, then matrix A has eigenlanguage $\langle \lambda_P, P^T A^k \rangle$ or $\langle \lambda_S, A^k S \rangle$, and matrix A^k has eigenlanguage $\langle (\lambda_P)^k, P \rangle$ or $\langle (\lambda_S)^k, S \rangle$.

Now consider the connection of these results with the theory of context-free (CF) grammars.

According to the results of (Kuich and Salomaa, 1986; Tarakanov, 1988), any CF-grammar $\langle G \rangle$ can be represented as a vector–matrix equation in one of the following forms:

$$Z = L(Z)Z + \overline{Z}, \qquad (4.16)$$

$$Z^T = Z^T R(Z) + \overline{Z}^T, \qquad (4.17)$$

where $Z = [z_1\ \ldots\ z_n]^T$ is a vector of nonterminals, or *syntactic classes* of the grammar, $\overline{Z} = [\overline{z}_1 \ldots \overline{z}_n]^T$ is a vector of terminals, and $L(Z)$, $R(Z)$ are square matrices over a semiring of nonterminals. It appears that matrix eigenlanguages allow for the determination of some properties of a language $\Im(\langle G \rangle)$ generated by the grammar directly from the matrices $L(Z)$, $R(Z)$, without generating any terminal chains.

Proposition 4.5 If a matrix $L(Z)$ has a right eigenlanguage $\langle \lambda_X, Z \rangle$, then $Z = \lambda_X^* \overline{Z}$, and any word of the language $\Im(\langle G \rangle)$ includes a root from the set λ_X^* and a terminate from the set $\overline{Z}$. If the matrix $R(Z)$ has a left eigenlanguage $\langle \lambda_Y, Z \rangle$, then $Z = \overline{Z}^T \lambda_Y^*$, and any word of the language $\Im(\langle G \rangle)$ includes a root from the set λ_Y^* and a prefix from the set $\overline{Z}$.

Proposition 4.6 If the matrix $L(Z)$ has a prefix eigenlanguage $\langle \lambda_P, P = [p_1 \ldots p_n]^T \rangle$, then the set $\Im_P = \bigcup_i p_i z_i$ of syntactic classes z_i with prefixes p_i is a language of type $\Im_P = \lambda_P^* P^T \overline{Z}$. If the matrix $R(Z)$ has a suffix eigenlanguage $\langle \lambda_S, S = [s_1 \ldots s_n]^T \rangle$, then a set $\Im_S = \bigcup_i z_i s_i$ of syntactic classes z_i with suffixes s_i is a language of type $\Im_S = \overline{Z}^T S \lambda_S^*$.

Example 4.13 Consider a CF-grammar without a chosen initial symbol and with the following rules:

$z_1 \to unsz_1$, $z_1 \to umusuz_2$, $z_1 \to un$, $z_2 \to msz_1$, $z_2 \to nsuz_2$, $z_2 \to m$,

where z_1, z_2 are nonterminals, and the letters m, n, s, u are terminals. For this grammar

$$L(Z) = \begin{bmatrix} uns & umsu \\ ms & nsu \end{bmatrix},$$

$$Z = \begin{bmatrix} z_1 \\ z_2 \end{bmatrix}, \overline{Z} = \begin{bmatrix} un \\ m \end{bmatrix}.$$

Such a matrix has a prefix eigenlanguage

$$\langle \lambda_P = sun + sum, P = [s, su]^T \rangle,$$

and therefore, according to Proposition 4.6, we obtain

$$\lambda_P^* P^T \overline{Z} = (sun + sum)^+.$$

78 Language Representation

Thus, the conjunction of the languages $sz_1 \cup suz_2$ forms the language consisting of the words *sun* and *sum*.

These results are explicit in addressing the properties of matrices over the field of real or complex numbers, and exhibit the potential for some useful applications of the IC approach to language representation and problem solving for at least the following reasons.

First, the results concerning degenerate and incomplete matrices coincide with those for real or complex matrices. But for complete matrices, the notions of prefix and suffix eigenlanguages have no analogies for real or complex matrices, since they represent specific corollaries of the noncommutativity of elements of semirings.

Second, any real or complex matrix of dimension $n \times n$ always has exactly n eigenvalues (taking into account their multiplicity) over the field of complex numbers (Gantmakher, 1988). However, not every matrix over a semiring has even one eigenlanguage. At the same time, not all matrices of real numbers have real eigenvalues.

Third, the essential distinction of a matrix over a semiring is that the concrete decomposition can determine only one eigenvector and eigenelement. At the same time the spectral decomposition of real or complex matrices is determined by a complete set of eigenvalues and eigenvectors, whose number is connected to the rank of the matrix.

4.3 Knowledge-Based Reasoning

Consider a set of rules of formal *T-cells* from Section 2.3.2 of the following form:

$$P_{k_0} \to \langle T_k \rangle P_{k_1} ... P_{k_n}. \qquad (4.18)$$

Add also *T-cells* of two specific types. Assume that we have an "initial" rule for some type k_i,

$$P_0 \to P_{k_i}, \qquad (4.19)$$

and a set of "terminal" rules for some of the types $k_i,...,k_j$,

$$P_{k_i} \to p_{k_i},...,P_{k_j} \to p_{k_j}. \qquad (4.20)$$

According to (Tarakanov and Dasgupta, 2000), P_0 can be regarded as corresponding to an antigen, while rules (4.20) correspond to *T-cells* that synthesize FPs independently from binding with any FP.

According to (Tarakanov, 1999a), a set of *T-cells* described by rules (4.18) through (4.20) is equivalent to a special kind of attributive CF grammar, where antigen P_0 corresponds to the axiom of the grammar, $P_{k_1}, \ldots, P_{k_n}$ correspond to nonterminals, p_{k_i}, p_{k_j} correspond to terminals, and symbols in the angle brackets correspond to synthesized attributes. This method can produce some grammars for solving tasks generalized as an inference engine.

Example 4.14 Consider triangle ABC in Figure 4.1 with angles A, B, C, and sides a, b, c.

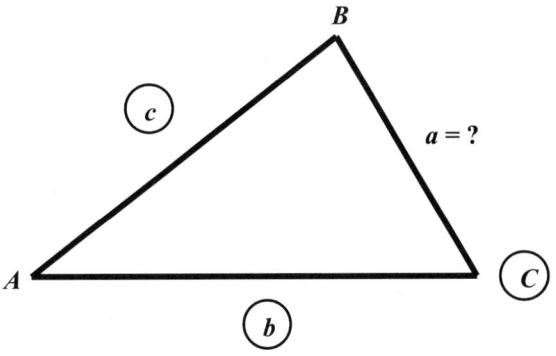

Figure 4.1 Triangle.

It is known that the parameters of any triangle satisfy the following equations:

$$A + B + C = \pi \quad \text{(theorem of angles)},$$

$$\frac{a}{\sin A} = \frac{b}{\sin B} = \frac{c}{\sin C} \quad \text{(law of sines)},$$

$$a^2 = b^2 + c^2 - 2(bc)\cos A \quad \text{(law of cosines)},$$
$$b^2 = a^2 + c^2 - 2(ac)\cos B \quad \text{(law of cosines)},$$
$$c^2 = a^2 + b^2 - 2(ab)\cos C \quad \text{(law of cosines)}.$$

A *T-cell* model of the triangle could be the following:

$P_1 = A$, $P_2 = B$, $P_3 = C$, $P_4 = a$, $P_5 = b$, $P_6 = c$,
$T_1 = T\mathrm{ang}$, $T_2 = T\mathrm{sin}$, $T_3 = T\mathrm{cos}$,
$P_1 \to \langle T_1 \rangle P_2 P_3$, $P_1 \to \langle T_2 \rangle P_2 P_4 P_5$,
$P_2 \to \langle T_1 \rangle P_1 P_3$, $P_2 \to \langle T_2 \rangle P_3 P_5 P_6$,
$P_3 \to \langle T_1 \rangle P_1 P_2$, $P_3 \to \langle T_2 \rangle P_1 P_6 P_4$,
$P_4 \to \langle T_3 \rangle P_1 P_5 P_6$, $P_5 \to \langle T_3 \rangle P_2 P_4 P_6$, $P_6 \to \langle T_3 \rangle P_3 P_4 P_5$.

These rules can also be considered as a knowledge base about triangles.

Consider the following task: Find a when C, b, c (circled in Figure 4.1), are given:

$$P_0 \to P_4, \quad (4.21)$$
$$P_3 \to P_3,$$
$$P_5 \to P_5, \quad (4.22)$$
$$P_6 \to P_6.$$

This task can be solved in the following way. First, given FPs that correspond to the rules (4.22), activate a *T-cell* that synthesizes an FP of the type P_2:

$$P_2 \to \langle T_2 \rangle P_3 P_5 P_6.$$

Second, this FP binds a receptor of the same type, and together with given P_3, activates the corresponding *T-cell* that synthesizes P_1:

$$P_1 \to \langle T_1 \rangle P_2 P_3.$$

Third, this FP together with given P_5 and P_6 activates the corresponding *T-cell* that synthesizes P_4:

$$P_4 \to \langle T_3 \rangle P_1 P_5 P_6.$$

Finally, P_4 activates a *T-cell* of rule (4.21), which results in the following solution of the task:

$$P_0 \to \langle T_3 \rangle \langle T_1 \rangle \langle T_2 \rangle P_3 P_5 P_6 P_3 P_5 P_6,$$

which usually implies that

$$a \to \langle T\cos \rangle \langle T\mathrm{ang} \rangle \langle T\sin \rangle CbcCbc.$$

Thus, the following solution has been synthesized:

1. Find angle B given angle C and sides b and c using the law of sines;
2. Find angle A by known angles B and C using the theorem of angles;
3. Find side a by known angle A and sides b and c using the law of cosines.

Moreover, the solution is synthesized in the so-called "prefix Polish notation," which can be interpreted strictly algorithmically.

Although this geometric example is rather simple, it shows the general principles of knowledge-based reasoning by IC. It also shows that FIN with *T-cells* represents a kind of inference engine for problem solving by simulation behavior of immune networks.

4.4 Linguistic Binding

It is known that natural proteins represent chains (or words of an alphabet) of 20 amino acids. Usually, no attention is paid to the fact that the number of amino acids is approximately equal to the number of letters in the alphabets of the so-called classical Indo-European languages. For example, the Italian alphabet has 21 letters, the Greek alphabet has 24 letters, and the English alphabet has 26 letters. But this analogy suggests that the concept of FP could unite the biological and symbolic (see Section 1.1) within the framework of a unified mathematical model.

Consider an alphabet and connect to each letter an FP with fixed coefficients of a quadratic form, just as in Section 4.1. Then every word over this alphabet will determine the concrete value of the free energy of the corresponding FPs. According to Definition 4.1, suppose that the value of the energy, which is higher than some threshold, determines the instability and decay of corresponding FPs. Thus we have in general an instrument for the formation of "correct words," or morphology. Furthermore, determine the binding energy between FPs in bilinear form, in accordance with Definition 2.4, and determine binding as in Definition 2.5. Then we have an instrument for the formation of "correct sentences," or grammar.

It is necessary to note that linguistic models of the natural languages are consistent with this approach. This can be seen in *the theory of linguistic valence* (TLV) by the French linguist L. Tesnière (Tesnière, 1965). This theory, however, occupies a somewhat isolated position in linguistics, since it strongly differs from the widely known generative grammars of N. Chomsky (Ginsburg, 1966). This allows us to detect at least the two following facts.

First, the TLV theory is tightly connected to a theory of *case* (or *role*) *grammars*. This theory describes the semantics of sentences as a system of *semantic valences*, which is determined by connections of *principal verbs* with other parts of a sentence. Furthermore, the roles of these parts are also determined by the principal verb. Within the framework of case grammar, several formal languages have been described at a semantic level, and the results are widely used in AI.

Second, nothing similar to Chomsky's "inherent grammar" has yet been discovered among biological structures. Moreover, even the existence of such a grammar seems rather problematic. At the same time, TLV considers the ability of a word to enter into syntactic connections with other elements on the basis of direct chemical analogy, and even fixes it in the title.

Tesnière claimed that a verb can be imagined as an original *atom with hooks*, which can attract to itself a greater or smaller number of actants depending on the greater or smaller number of hooks that it has to hold these actants to itself. The number of such hooks available for a verb, and therefore the number of actants that it is capable of controlling, is the essence of the term *valence of a verb*. So, linguistic valence is the ability of a word to enter into syntactic connections with other words of the language. Tesnière, utilized this term as a label of compatibility, using it only for verbs.

It is worth noting that the developing IC approach can also be used for the mathematical formalization of TLV. Instead of "atom with hooks," each verb could be seen as an FP with secondary structures (sub-FPs), whose number is equal to the valence of the verb. In this way, using the notion of binding (see Definition 2.5) we can determine a general tool for generation and selection of syntactically and/or semantically correct sentences.

In this connection it is necessary to make some improvements concerning the term *valence*. In chemistry, valence concerns the so-called *strong* interactions, whose energy is much higher than the so-called *weak*, or nonvalence, interactions between proteins (see Section 1.2). Therefore, within the IC approach we would propose, instead of the term "linguistic valence," the term *linguistic binding*.

5
Modeling of Natural and Technical Systems

5.1 Spatial Structures of Native Proteins

This section uses the idea of a formal protein to model the parameters of its biological prototypes. In Section 5.1.1 we consider an algebraic model, which allows us to determine analytically the spatial parameters of the so-called *secondary structures* of native proteins. Then, in Section 5.1.2 we propose an approach to modeling the spatial shape of a protein defined by its amino acid sequence.

5.1.1 Algebraic Description of Secondary Structures

Currently, molecular biology has established experimentally and has provided theoretical explanations for some of the mechanisms of the secondary structures of proteins. In these structures, the most stable configurations of proteins and the spatial organization of the polypeptide skeleton are repeated. Such structures are formed and stabilized by hydrogen bonds, which can be found between groups of atoms **O–H...O** in the polypeptide skeleton (see Figure 2.1). The main secondary structures are *α-helix*, *β-sheet*, and *β-turn* (Bochinsky, 1987; Cantor and Schimmel, 1980a). It is thought (Shaitan, 1994) that the α-helix and β-sheet reinforce the protein molecule, while a β-turn allows a polypeptide chain to sharply change direction without losing stability.

Consider an analytic model of the parameters of secondary structures of proteins as a special case of FP.

Introduce two unitary quaternions:

$$Q_L(\varphi) = (q_1, q_2, q_3, q_4), \quad Q_R(\psi) = (p_1, p_2, p_3, p_4),$$

where

$$q_1 = -\sin\frac{\varphi}{2}\cos\frac{\theta+\eta}{2}, \quad q_2 = \cos\frac{\varphi}{2}\cos\frac{\theta-\eta}{2},$$

$$q_3 = -\cos\frac{\varphi}{2}\sin\frac{\theta-\eta}{2}, \quad q_4 = \sin\frac{\varphi}{2}\sin\frac{\theta+\eta}{2},$$

$$p_1 = \cos\frac{\psi}{2}\cos\frac{\theta-\eta}{2}, \quad p_2 = \sin\frac{\psi}{2}\cos\frac{\theta+\eta}{2},$$

$$p_3 = \sin\frac{\psi}{2}\sin\frac{\theta+\eta}{2}, \quad p_4 = -\cos\frac{\psi}{2}\sin\frac{\theta-\eta}{2},$$

According to (2.1), these quaternions correspond to the following matrices of dimension 3×3, where constants c_0, s_0, c_1, s_1 correspond to the designations of Section 2.1.1:

$$L(\varphi) = \begin{bmatrix} c_0 c_1 + s_0 s_1 \cos\varphi & s_0 c_1 - c_0 s_1 \cos\varphi & s_1 \sin\varphi \\ -c_0 s_1 + s_0 c_1 \cos\varphi & -s_0 s_1 - c_0 c_1 \cos\varphi & c_1 \sin\varphi \\ s_0 \sin\varphi & -c_0 \sin\varphi & -\cos\varphi \end{bmatrix},$$

$$R^T(\psi) = \begin{bmatrix} c_0 c_1 + s_0 s_1 \cos\psi & -c_0 s_1 + s_0 c_1 \cos\psi & -s_0 \sin\psi \\ -s_0 c_1 + c_0 s_1 \cos\psi & s_0 s_1 + c_0 c_1 \cos\psi & -c_0 \sin\psi \\ s_1 \sin\psi & c_1 \sin\psi & \cos\psi \end{bmatrix}.$$

Multiplication of these quaternions gives $Q_L(\varphi)Q_R(\psi)=Q(\varphi,\psi)$, where $Q(\varphi,\psi)$ is just a quaternion of the kind (2.1)–(2.2) from Definition 2.1 of FP. This quaternion corresponds to the so-called *torsion matrix* (Cantor and Schimmel, 1980a),

$$T(\varphi,\psi)=L(\varphi)R^T(\psi), \qquad (5.1)$$

which describes spatial rotations of links of the polypeptide skeleton of proteins (Figure 2.1).

Consider the quaternion $Q_M(\varphi,\psi)=Q_R(\psi)Q_L(\varphi)$, which can be called a mirror quaternion, because of the permutation of the multipliers related to the quaternion $Q(\varphi,\psi)$. Direct multiplication gives the following components of Q_M:

$$Q_M(\varphi+\psi) = \left(-c_0 \sin\frac{\varphi+\psi}{2}, \ \cos\frac{\varphi+\psi}{2}, \ 0, \ s_0 \sin\frac{\varphi+\psi}{2}\right). \qquad (5.2)$$

Consider a matrix of dimension 3×3, which corresponds to the quaternion (5.2), where $\sigma = \varphi+\psi$:

5.1 Spatial Structures of Native Proteins

$$M(\varphi+\psi) = \begin{bmatrix} c_0^2 + s_0^2 \cos\sigma & s_0 c_0(1-\cos\sigma) & s_0 \sin\sigma \\ s_0 c_0(-1+\cos\sigma) & -s_0^2 - c_0^2 \cos\sigma & c_0 \sin\sigma \\ s_0 \sin\sigma & -c_0 \sin\sigma & -\cos\sigma \end{bmatrix}. \quad (5.3)$$

Definition 5.1 Call the matrix $M(\varphi+\psi)$ of form (5.3), which corresponds to the quaternion $Q_M(\varphi+\psi)$ of form (5.2), a *mirror matrix*.

Speaking strictly, the mirror matrix and torsion matrix for the same torsion angles are congruent in terms of matrix theory (Horn and Johnson, 1986) or conjugate in terms of group theory (Olshansky, 1989), because the following relations are satisfied:

$$T(\varphi,\psi) = R(\psi)M(\varphi+\psi)R^T(\psi) = L(\varphi)M(\varphi+\psi)L^T(\varphi), \quad (5.4)$$

$$Q(\varphi,\psi) = Q_R^{-1}(\psi)Q_M(\varphi+\psi)Q_R(\psi) = Q_L(\varphi)Q_M(\varphi+\psi)Q_L^{-1}(\varphi). \quad (5.5)$$

Hence, these matrices determine the same linear transformation of 3D space, but as represented in different coordinate systems. At the same time, the mirror matrix (5.3) is much simpler than the initial torsion matrix. This can be checked by direct multiplication of matrices (5.1) or by comparing quaternions $Q_M(\varphi+\psi)$ of form (5.1) and $Q(\varphi,\psi)$ of form (2.5).

The spectral decomposition of a matrix by eigenvalues and eigenvectors, which are invariants of the linear transformation defined by the matrix, plays an important role in matrix theory and applications (Horn and Johnson, 1986). The rather simple appearance of the mirror matrix allows for an analytical determination of its spectral decomposition as follows:

$$M(\varphi+\psi) = M(\gamma,\phi) = F(\phi)R_X(\gamma)F^T(\phi), \quad (5.6)$$

where

$$R_X(\gamma) = \begin{bmatrix} 1 & 0 & 0 \\ 0 & \cos\gamma & \sin\gamma \\ 0 & -\sin\gamma & \cos\gamma \end{bmatrix}, \quad F(\phi) = \begin{bmatrix} \cos\phi & \sin\phi & 0 \\ 0 & 0 & 1 \\ \sin\phi & -\cos\phi & 0 \end{bmatrix},$$

$$\cos\gamma = -s_0^2 - c_0^2 \cos\sigma,$$
$$\cos\phi = \frac{c_0 \sin\sigma}{\sin\gamma}, \quad (5.7)$$
$$\sin\phi = s_0 c_0 \frac{1-\cos\sigma}{\sin\gamma}.$$

86 Modeling of Natural and Technical Systems

Thus, eigenvalues of the mirror matrix form a matrix $R_X(\gamma)$, while eigenvectors F_1, F_2, F_3 are columns of a matrix $F(\phi)=[F_1\ F_2\ F_3]$, where $F(\phi)F^T(\phi) = E$.

Definition 5.2 The angle γ is called a *twisting angle*, and angle ϕ an *eigenangle*, of the mirror matrix.

A geometric interpretation of these angles is given in Figure 5.1. Thus a mirror matrix rotates an arbitrary vector on the twisting angle around an axis F_1 (Figure 5.1a), whose position is determined entirely by the eigenangle (Figure 5.1b).

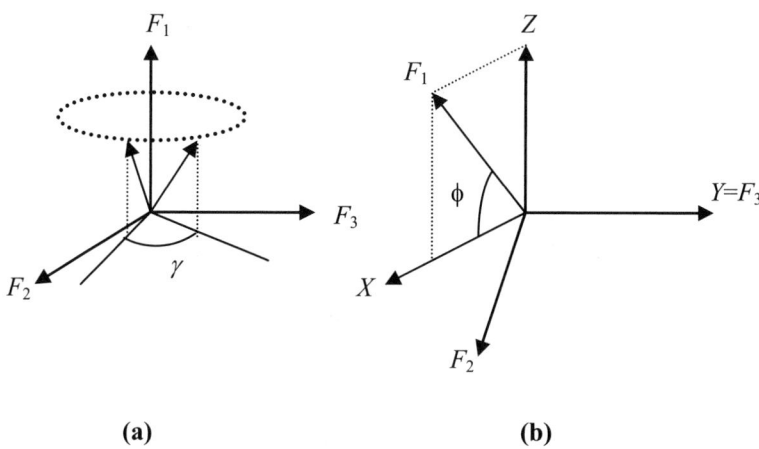

Figure 5.1 Geometric interpretation of twisting angle and eigenangle of a mirror matrix.

Substitution of (5.6) by (5.4) also allows us to determine the spectral decomposition of the torsion matrix. It is a given that eigenvalues of mirror and torsion matrices are the same, while eigenvectors of the torsion matrix are columns of a matrix $L(\varphi)F(\phi)=R(\psi)F(\phi)$. It is implied that the torsion matrix also rotates an arbitrary vector on the twisting angle γ. However, the position of the axis of rotation has a more complicated dependence on torsion and valence angles and does not have such an obvious geometric sense as does the mirror matrix.

Definition 5.3 A typical configuration of the polypeptide skeleton is called *n-helix*, where $\varphi_i=\varphi_j$, $\psi_i=\psi_j$ for every i,j, and the axes of torsion rotations of the residues i and $n+i$ are coplanar and parallel.

Note that in the sense of this definition a β-sheet can be treated as a 2-helix.

The above parameters of the mirror matrix allow us to describe the secondary structures of proteins in the following elegant form:

Theorem 5.1 The following conditions are necessary for the secondary structures of proteins:

$$\text{β-sheet:} \quad \cos(2\gamma) = 1; \qquad (5.8)$$
$$n\text{-helix:} \quad \cos(n\gamma) = 1; \qquad (5.9)$$
$$\text{β-turn:} \quad \phi=\gamma. \qquad (5.10)$$

We omit the proof of the theorem for economy of space (see Tarakanov, 1999a).

It is worth noting that the above conditions are necessary but not sufficient, because they describe any regular structure of a protein together with the secondary structures, which are stabilized by hydrogen bonds.

This discussion shows that the invariants of the mirror matrix describe β-turn (5.10) as something different from the other secondary structures (5.8), (5.9). However, an algebraic construction exists that allows us to determine a formal connection between a β-turn and a β-sheet.

According to (5.6), we can represent the SVD of the mirror matrix in the form $M=AEB^T$, where E is the identity matrix of the singular values, while A and B are matrices of left and right singular vectors:

$$A = F(\phi)R_X\left(\frac{\gamma}{2}\right), \quad B^T = R_X\left(\frac{\gamma}{2}\right)F^T(\phi). \qquad (5.11)$$

Using this SVD, introduce a *commutator* (Olshansky, 1989) of the mirror matrix, by which we will understand the commutator of matrices of singular vectors:

$$K=AB^TA^TB. \qquad (5.12)$$

The spectral decomposition of the commutator of the mirror matrix allows us to state that the twisting angle κ of the commutator is as follows:

$$\cos\kappa = \frac{1}{2}(1+\cos\gamma+\cos\phi-\cos\gamma\cos\phi)^2 -1. \qquad (5.13)$$

Theorem 5.2 The following conditions are necessary for the secondary structures of proteins:

$$\beta\text{-structure: } \kappa = \phi,$$

including

$$\beta\text{-sheet: } \kappa = \phi = 0,$$
$$\beta\text{-turn: } \kappa = \phi = \gamma.$$

We omit the proof of the theorem for economy of space (see Tarakanov, 1999a).

It is worth noting that the commutator formally establishes the existence of mirror symmetry in the β-structures. For example, the commutator of the β-sheet is a unit matrix, while for the β-turn some elements of the commutator $K=[k_{ij}]$, $i,j=1,2,3$, are permutable, and the structure of the commutator appears to share some features of the unit matrix:

$$k_{11}=k_{22},$$
$$k_{31}=k_{23},$$
$$k_{13}=k_{32},$$
$$k_{12}+k_{21}+k_{33}=1.$$

Numerical verification of the obtained analytical results obtained was carried out using an FP with fixed valence angles $\theta=60°$, $\eta=40°$.

According to (5.7)–(5.9), the sums of the torsion angles for the *n*-helices, where $n=2,3,4$, have been computed by the formula

$$\cos(\varphi+\psi) = -\frac{s_0^2 + \cos\gamma}{c_0^2},$$

$$\gamma = \frac{2\pi}{n}k,$$

$$k = 0,1,\ldots,n-1.$$

The results of the computation are shown in Table 5.1, which also contains experimental data from (Cantor and Schimmel, 1980a) for comparison.

For the β-turn, computing of the torsion angles gives two variants:

1. $\varphi_1 = \psi_2 \cong 203°$, $\psi_1 \cong 84°$, $\varphi_2 \cong 72°$;
2. $\varphi_1 = \psi_2 \cong 157°$, $\psi_1 \cong -84°$, $\varphi_2 \cong -72°$.

According to a computer model of the spatial configurations of the polypeptide skeleton, these variants correspond to the variants given in (Cantor and Schimmel, 1980a). It is worth noting that the hydrogen bonds for the computed variants are strictly collinear.

Secondary structure	$\cos(\gamma)$	$\cos(\varphi+\psi)$	$\varphi+\psi$ Computed	$\varphi+\psi$ Experimental
β-sheet (2-helix)	−1	1	0°	−6° : parallel β-sheet −4° : antiparallel β-sheet (β-polyalanine)
3-helix	$-\dfrac{1}{2}$	$\dfrac{1}{3}$	±70.5°	70° : polyglycine 75° : polyproline-1 71° : polyproline -2
α-helix (4-helix)	0	$-\dfrac{1}{3}$	±109.5°	−104° : right α-helix (α-polyalanine) 104° : left α-helix

Table 5.1 Computed and experimental parameters of secondary structures of proteins.

Based on the model of FP, the mirror matrix is similar to the torsion matrix, having, however, a much simpler form. The parameters of the spectral decomposition of the mirror matrix and its commutator determine the main secondary structures of proteins and give compact analytical formulae for approximate estimation of the values of the torsion angles. According to the computer model of the spatial conformations of the polypeptide skeleton, these values have good correspondence (error of no more than 6°) with the experimental data. Apparently, a source of the errors is in the accepted assumptions of the valence angles $\eta=\xi$ for the FPs.

5.1.2 Dependence of the Configuration on the Amino Acid Sequence

There are some rather convincing proofs that the spatial configuration and physiological properties of proteins are determined, in fact, by their amino acid sequence (e.g., see Bochinsky, 1987). Therefore, the problem of determining spatial configuration by the amino acid sequence of a protein is of great importance for molecular biology

For example, in the work of (Kuzminov, 1987) such dependence is studied in detail for relatively short polypeptides. For this purpose a conformation analysis has been carried out for tetrapeptides, which include the same amino acid residues: Ala – alanine, Leu – leucine, Phe – phenylalanine, Trp – tryptophan. The following three chains of tetrapeptides have been studied, which include the given amino-acids in different sequences:

1. Chain 1:   PheTrpLeuAla;
2. Chain 2:   LeuAlaPheTrp;
3. Chain 3:   TrpLeuPheAla.

90 Modeling of Natural and Technical Systems

In addition, for every chain all eight possible forms of the polypeptide structure have been investigated depending on the relative positions of side groups.

The results of the study are shown in Table 5.2, which contains the values of the free energy for the optimal conformation of every chain in every possible form. These values have been computed in (Kuzminov, 1987) by the atomic potential method, using minimization of the energy of nonvalence and electrostatic interactions, as well as hydrogen bonds and frozen rotation around the valence bonds.

Consider an IC interpretation of these results from the viewpoint of FP.

Assume four monopeptides, Ala, Leu, Phe, Trp, which compose three tetrapeptides of the sequence of chains 1 through 3. Assume that every monopeptide can be in only one of two states (forms), which we will denote by the letters a and b. Then the form of any tetrapeptide can be coded by a four-letter chain in the alphabet $\{a, b\}$. Choose the first letter arbitrarily. Then all possible eight forms of the tetrapeptides studied by (Kuzminov, 1987) can be coded by combinations of second, third, and fourth letters, as shown in Table 5.2.

Conformation Chain	bbbb	bbba	bbaa	baaa	baab	babb	aaba	abab
PheTrpLeuAla	35.8	36.9	38.2	42.8	36.9	32.9	36.6	38.2
LeuAlaPheTrp	38.9	40.5	41.0	40.5	36.6	34.8	39.8	37.5
TrpLeuPheAla	38.5	39.0	36.4	40.0	36.8	34.0	40.8	36.2

Table 5.2 Free energy ($-v$ kcal/mol) for a tetrapeptide chain in different conformations.

Consider the following task: to determine the controls of an FP and the states of four monopeptides, which can be in two forms each, so that the values of the free energy of the FP correspond to those in Table 5.2.

This task has been solved by computer modeling of the FP with fixed valence angles $\theta=60°$, $\eta=40°$. We used a stochastic search (Monte Carlo method) to find eight pairs of torsion angles φ, ψ (four monopeptides in two forms each) and four controls v_{11}, v_{22}, v_{33}, v_{44} (all other controls are considered equal to zero for simplification). The goal of the search was the minimization of the sum of squared deviations of computed energy values from their values shown in Table 5.2 for all chains in all conformations. The results of the computations are shown in Table 5.3, where $\max|\Delta v^{(i,j)}|$ means a module of the maximal (over all conformations) deviation of the energy of the ith chain from the value given in Table 5.2.

The fourth column of Table 5.3 contains states of the monopeptides that minimize the deviation of energy of the FP from the values of Table 5.2 over all chains in all conformations. These results can also be treated as an approximation of the given set of numbers (Table 5.2) by the parameters of the FP. Note that the approximation error does not exceed 1.5%.

Suppose, now, that the obtained configurations of the monopeptides are slightly perturbed by being included in different chains. This supposition allows us to decrease up to ten times the approximation errors for a concrete chain, as shown in the last three columns of Table 5.3. At the same time, the deformation of the torsion angles does not exceed 10°, and the maximal errors for the chains, whose energy was minimized, are underlined in Table 5.3.

It is interesting to note that the computed values of the torsion angles are disposed in the fields set experimentally for the monopeptides of the alanine type (Cantor and Schimmel, 1980a, Kuzminov, 1987). For example, amino acid residues of α-polyalanine are arranged in the field that corresponds to the conformation a: ($\varphi=-57°$, $\psi=-47°$), while the residues of β-polyalanine are arranged in the field b: ($\varphi=-139°$, $\psi=135°$).

Thus, the computations show good correspondence with the experimental data.

Peptide (form) and Chains	Parameters of FP and max deviations of energy	minimum for all chains	minimum for Chain 1	minimum for Chain 2	minimum for Chain 3
Phe (b)	$\varphi°, \psi°$	−151, 123	−154, 131	−144, 129	−146, 127
Phe (a)	$\varphi°, \psi°$	−151, 41	−152, 44	−153, 42	−157, 39
Trp (b)	$\varphi°, \psi°$	−154, 110	−155, 110	−156, 109	−160, 112
Trp (a)	$\varphi°, \psi°$	−100, 9	−102, 7	−101, 10	−112, 0
Leu (b)	$\varphi°, \psi°$	−166, 103	−167, 109	−167, 104	−170, 107
Leu (a)	$\varphi°, \psi°$	−74, −60	−75, −58	−70, −59	−68, −52
Ala (b)	$\varphi°, \psi°$	−179, 96	−180, 96	−180, 94	−180, 95
Ala (a)	$\varphi°, \psi°$	−38, −70	−41, −71	−39, −66	−43, −76
Chain-1: PheTrpLeuAla	max$|\Delta v^{(1,j)}|$	0.40	<u>0.04</u>	1.90	1.60
Chain-2: LeuAlaPheTrp	max$|\Delta v^{(2,j)}|$	0.40	0.60	<u>0.04</u>	2.00
Chain-3: TrpLeuPheAla	max$|\Delta v^{(3,j)}|$	0.40	0.63	0.80	<u>0.05</u>

Table 5.3 States of monopeptides for the controls $v_{11}=36.0$, $v_{22}=39.2$, $v_{33}=29.5$, $v_{44}=51.1$, and maximal deviations of energy of the tetrapeptidic FPs (kcal/mol) from the values in Table 5.2.

5.2 Synchronization of Events in Computer Networks

Consider an IC model of synchronization of events in distributed asynchronous systems. The model uses principles of homology and recognition between FPs. As a result, the model does not need to introduce any notion of "time," which is usual for algorithms that provide the synchronization of events or the necessary order of delivering messages, and that are called *synchronization protocols* (Braudes and Zabele, 1993). Two such well-known protocols, based on *scalar time* and *vector time*, turn out to be specific cases of the IC model. Therefore, the model seems to be useful for high-speed networking and open distributed processing (ODP) as a general approach to flexible synchronization of events with numerous special requirements (Gorodetski and Tarakanov, 1995).

5.2.1 Time-Based Multicast Protocols

Multicast or one-to-many communications within a distributed group is becoming popular as the number of applications using audio and video data streams increases. Today's multicast applications are primarily local and of small scale. Factors such as protocol simplicity and convenience of development have dominated protocol design activities, and experience demonstrates that current protocols do not address all of the requirements of multicast applications (Braudes and Zabele, 1993). Moreover, new requirements arise in the context of wide-area high-speed networking on ODP platforms, where clients and servers with heterogeneous resources would be allowed real-time join and leave (De Meer et al., 1994).

One of the central problems of multicast is distributed synchronizing, or deciding how to provide an order for message delivery. Various approaches to ordering messages in asynchronous systems have been studied. Lamport (1978) proposed logical clocks to produce a total ordering on messages. Birman et al. (1991) presented several types of vector clocks. Their ideas have been realized in well-known multicast protocols, such as the *Atomic Broadcasting Protocol Based on Lamport Time Stamps* (ABCAST LT) and the *Causal Broadcasting Protocol Based on Vector Time Stamps*) (CBCAST VTS). However, any similar protocol is effective only under rather restricted conditions (e.g., the intensity of the message stream). Therefore, different conditions require special protocols, which are based on their unique models.

Remarkably, similar problems are solved on the level of biomolecules where billions of heterogeneous cells have to synchronize their functionalities by exchanging proteins as "multicast messages." On this level there is nothing similar to scalar or vector clocks, but we can find rather general and effective mechanisms of synchronizing (e.g., see Alberts et al., 1986), and these are the subject of formalization and modeling (using the IC approach) in this section.

The remainder of Section 5.2 is organized as follows. In Section 5.2.2 we propose a multicast model in the terms of FP homology. Then in Section 5.2.3

we consider the ABCAST LT and CBCAST VTS protocols as special cases of the model. These considerations underline the self-organizing feature of the general model without introducing any kind of "time." In Section 5.2.4 we sketch a formal description of the model in terms of a special FIN with *T-cells*. In conclusion, we explain why the IC multicast model could be useful for computer networks applications.

5.2.2 IC Model of Multicast Protocols

The main features of biomolecular mechanisms are determined by their distributed and self-organizing nature. There is no local center that could control billions of cells simultaneously by observing their states and sending corresponding messages to every cell. In spite of this, cells exchange proteins as messengers or biomolecular signals, and the cells are thus cooperative and self-synchronizing.

Some of the fundamental mechanisms of such complex behavior have been understood for some time, including molecular recognition based on complementarity (homology) and high concurrence (Alberts et al., 1986). Since there is no need to discuss biomolecular details in this section (see Section 5.1), we now move on to the terms of multicast.

Consider a problem where computing proceeds by a sequence of broadcasts, in which a process sends a message to some arbitrary subset of processes, including itself.

Let some process P_{sen} send a message $\{M\}$ to a group of processes that include P_{rec}. Assume that P_{sen} has already received messages $\{A, B, C, D\}$ before it sends message $\{M\}$. Assume that P_{rec}, before receiving the message $\{M\}$, has already received the following messages in the order $\{A, C, B\}$, where the same letters designate the same messages. Let lowercase letters $\{a, b, c, d, m\}$ designate the codes (names) of the messages. Assume that P_{sen}, when sending message $\{M\}$, includes the code (chain) of the received messages, including the last one: $\langle \text{-}a\text{-}b\text{-}c\text{-}d\text{-}m\text{-}\rangle$. This chain replicates itself, and its copies are received by multicast processes. When receiving a copy, P_{rec} synthesizes its own "point of view" on the order of messages $\langle \text{-}a\text{-}c\text{-}b\text{-}m\text{-}\rangle$ and tries to compare it with the received one, as shown in Figure 5.2.

Let two chains interact, according to their homology, as shown in Figure 5.2b,c, where homologous codes are linked by vertical lines. As one can see from Figure 5.2, this interaction is able to expose locations (*loops*) where homology is violated. According to such loops, the main task of multicast protocol is reduced to deciding whether the synthesized chain of P_{rec} is consistent with the received chain of P_{sen}. In the case of inconsistency the received message should be delayed in an *input buffer*, while consistency indicates that the received message should be delivered to P_{rec}.

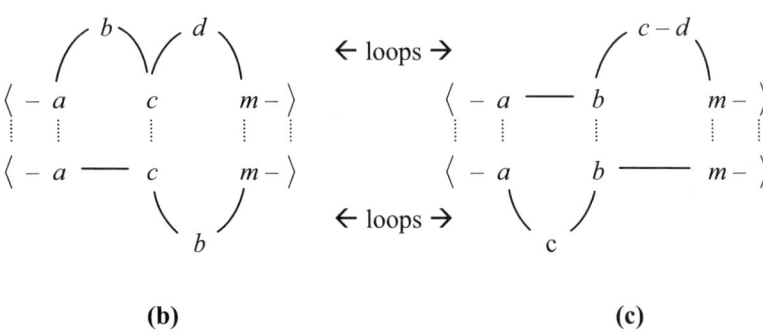

Figure 5.2 Homologies between chains of messages.

Let two chains interact, according to their homology, as shown in Figure 5.2b,c, where homologous codes are linked by vertical lines. As one can see from Figure 5.2, this interaction is able to expose locations (*loops*) where homology is violated. According to such loops, the main task of multicast protocol is reduced to deciding whether the synthesized chain of P_{rec} is consistent with the received chain of P_{sen}. In the case of inconsistency the received message should be delayed in an *input buffer*, while consistency indicates that the received message should be delivered to P_{rec}.

Although this model appears to be rather simple, it allows us to develop a number of multicast protocols for ordering message delivery. Within the general model we can change the mode of coding messages, the rules of destroying, the possibilities of chain transformation, including transpositions of letters, etc. Two corresponding examples are considered below.

5.2.3 Special Cases

Although the model does not use any kind of time, it is able to synchronize message delivery. Consider two well-known time-based protocols as special cases of the model.

Scalar Time Stamps

In his classic work, Lamport (1978) showed that the role of "time" in distributed systems consists in the ordering of events. How can one prescribe time to every event in a distributed system so that the order of events will be reasonable and correct? Evidently, this could be done by some "oracle" that knows the global physical time and can be accessed by each process of the system. However, realization of global time in a distributed system meets serious problems, because every node of the system has its own delays and intensity of events.

As an alternative to global physical time, Lamport proposed the use of *logical time* in distributed systems. The single role of logical time is ordering of events so that two events that have causal dependence are connected by a "before–after" relation within this artificial time.

The ABCAST LT protocol, based on Lamport time stamps, includes the following rules:

1. Before sending any message $\{M\}$ the process P_{sen} increases its discrete local time,

$$LT(P_{sen}) := LT(P_{sen}) + 1,$$

and stamps message $\{M\}$ with this value,

$$LT(M) := LT(P_{sen}).$$

2. After receiving any message $\{M\}$ the process P_{rec} installs its local time, so that

$$LT(P_{rec}) := \max\{LT(P_{rec}), LT(M)\}.$$

After that, P_{rec} delays delivering message $\{M\}$ until all other channels are *flushed up* to the time $LT(M)$.

3. If message $\{R\}$ has the minimal stamp among all received messages with all channels flushed up to the time $LT(M)$, then the message $\{R\}$ can be delivered to P_{rec}.

The first step of the protocol guarantees that the messages sent by P_{sen} have different and increasing stamps. The next step supports the logical clock of P_{rec} and enssures that the delay in delivery will not exceed the time when all messages with earlier stamps would be properly received by the input buffer. The last step delivers messages in increasing order of their stamps.

Now consider an IC model where all messages are coded by the same letter, e.g., by $\{m\}$. Then the time stamp of any message represents the length of the chain, i.e., the number of letters $\{m\}$ in the chain. From this viewpoint the steps of the ABCAST LT protocol can be treated as follows.

The first step consists in adding one letter $\{m\}$ to the chain of P_{sen}. In the next step P_{rec} synthesizes a chain with corresponding length. In the last step chains interact by pairs and destroy (delay) those that have loops.

An example of the last step is depicted in Figure 5.3 for a set of messages with time stamps $\{3,2,1\}$.

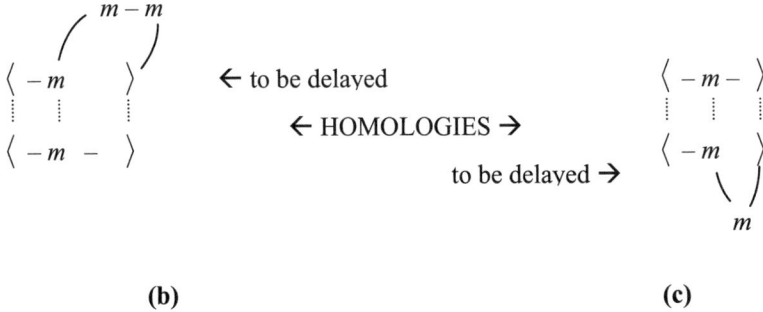

(a)

(b)   (c)

Figure 5.3 Homologies between messages with time stamps $\{3, 2, 1\}$.

Introducing the concepts logical clock and logical time stamp allows us to solve several multicast problems in a rather simple way. The ABCAST LT protocol is very useful in some applications with intensive and permanent streams of messages. But in general it needs considerable resources for channel flushing. Moreover, it orders messages that are not in causal order. Some of these drawbacks can be overcome by introducing vector time stamps.

Vector Time Stamps

Assume a multicast group of n processes. Then the vector time of some process $P_{rec.i}$ is a vector of length n whose component j represents a number of messages delivered to $P_{rec.i}$ from $P_{sen.j}$.

The CBCAST VTS protocol), based on vector time stamps, includes the following rules:

5.2 Synchronization of Events in Computer Networks

1. Before sending a message $\{M\}$ to all members of the group, process $P_{sen.j}$ increases the component j of its local vector clock,

$$VT(P_{sen.j})[j] := VT(P_{sen.j})[j] + 1,$$

and stamps the message with a vector stamp

$$VT(M) := VT(P_{sen.j})[j].$$

2. After receiving the message $\{M\}$ sent by $P_{sen.j}$ and stamped with the vector time stamp $VT(M)$, the receiving process $P_{rec.i}$ delays delivery of the message until the following conditions become true:

$$VT(M)[j] = VT(P_{rec.i})[j] + 1,$$
$$VT(P_{rec.i})[k] \geq VT(M)[k], \; k=1,\ldots,j-1, j+1,\ldots, n.$$

3. When message $\{M\}$ is delivered to $P_{rec.i}$, the local clock of the last process is installed as follows:

$$VT(P_{rec.i})[k] := \max\{VT(P_{rec.i})[k], VT(M)[k]\}, \; k=1,\ldots, n.$$

The first step of the protocol has the same sense as for the Lamport time stamps. The first condition of the second step means that $\{M\}$ is the next message in turn among those from $P_{sen.j}$ to $P_{rec.i}$, which implies that no single message is lost or delayed in the channels. The second condition of the second step guarantees that all messages received by $P_{sen.j}$ are also received by $P_{rec.i}$. The third step synchronizes the vector clock of $P_{rec.i}$ according to the vector clock of $P_{sen.j}$.

Now consider an IC model where messages sent by the same process are coded with the same letter, but the letters are different for different processes. Let a reading of the vector clock represent a number of corresponding letters in a chain and the order of different letters is the same for every chain (Figure 5.4 shows an example of such coding).

From the viewpoint of the IC model, the steps of the CBCAST VTS protocol) can be treated as follows. The first step consists in adding a definite letter to the corresponding subchain of P_{sen}. In the next step, the chain from P_{sen} interacts with the chain from P_{rec}, and if the first chain has a loop (as shown in Figure 5.4b), then it should be delayed in the input buffer of P_{rec}.

Using the vector clock in distributed systems has its own drawbacks, e.g., large memory for storing every process, the vector times of all multicast processes, and the great length of the time stamps, which can overflow the channel. Apparently, these drawbacks can be overcome by the IC model, because similar problems have already been solved via interactions within the natural immune networks. Consider a possible formal description of the model.

$$VT(P_{sen.3})=[2\ 1\ 1\ 1] \quad \langle -a-a-b-c-d- \rangle$$
$$VT(P_{rec.4})=[1\ 1\ 0\ 1] \quad \langle -a-b-d- \rangle$$

(a)

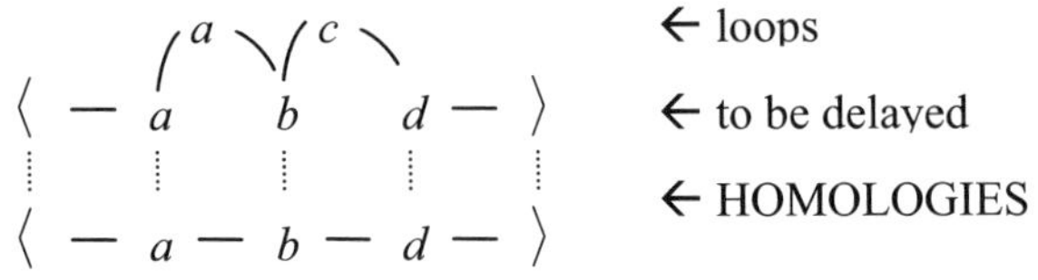

$$\langle - \overset{a}{\underset{a}{\diagdown}} \overset{c}{\underset{b}{\diagup}} d - \rangle \quad \leftarrow \text{loops}$$
$$\qquad\qquad\qquad\qquad \leftarrow \text{to be delayed}$$
$$\langle - a - b - d - \rangle \quad \leftarrow \text{HOMOLOGIES}$$

(b)

Figure 5.4 Example of homologies between messages [2111] and [1101].

5.2.4 Specification of the Model

A formal description of the IC model can be based on a special kind of FIN with *T-cells* (see Section 2.3.2) and knowledge-based reasoning by such an FIN (see Section 4.3). For this purpose define *T-FIN* such that it includes the following kinds of *T-cells*:

$$\langle T\text{-}FIN \rangle = \langle \{A, B, \ldots\}, \{a, b, \ldots\}, T_0, T_A, T_B, \ldots, T_a, T_b, \ldots \rangle,$$

where $\{A,B,\ldots\}$ and $\{a,b,\ldots\}$ are FPs, and T_0, T_A, $T_B,\ldots$, T_a, $T_b,\ldots$ are *T-cells*. Among these, *T-cells* T_A, T_B, etc., correspond to Definition 2.9, just as in (4.18), while the other *T-cells* can be treated as *reduced T-cells*, where T_0 corresponds to the initial rule, as in (4.19), and T_a, T_b, etc., correspond to terminal rules, as in (4.20).

Assume that any P_{rec} generates one of the terminal rules, e.g.,

$$A \to a,\ C \to c,\ B \to b,\ M \to m,\ \ldots,$$

every time it receives a message with a corresponding code. This set of rules does not form a grammar due to the lack of an initial rule. Assume that any P_{sen} sends just such an initial rule in its message to a multicast group. Let this rule correspond to the sequence of the messages received by P_{sen}, e.g., in Figure 5.2:

5.2 Synchronization of Events in Computer Networks

$$T_0 \to ABCDM.$$

Now we have a CF-grammar, and the problem of message delivery order to P_{rec} can be reduced to a question of whether any terminal string exists in this grammar.

However, grammar inference is not able to expose all loops. For example, in a grammar

$$T_0 \to ABC,$$
$$A \to a,$$
$$C \to c,$$
$$B \to b,$$

a terminal string *abc* exists, but the grammar is not able to expose the first upper loop in Figure 2b. However, in cases of scalar and vector clocks, all the relevant loops can be exposed in the following way.

Assume that rules are generated according to a value of a vector clock; e.g., in Figure 4a,

$VT(P_{sen}) = [2111]$:

$$T_0 \to A[1]A[2]B[1]C[1]D[1];$$

$VT(P_{rec}) = [1101]$:

$$A[1] \to a[1], \; B[1] \to b[1], \; D[1] \to d[1].$$

Here FPs $A[1]$, $A[2]$, ..., $B[1]$, etc., correspond to nonterminals, and FPs $a[1]$, $a[2]$,..., $b[1]$, etc., correspond to terminals. No terminal strings can be derived in this grammar, but all possible substitutions expose the loops that remained nonterminals, e.g., $A[2]$ and $C[1]$, during this interaction:

$$T_0 \to a[1]A[2]b[1]C[1]d[1].$$

Evidently, scalar time grammar can be considered as a special case of vector time grammar using only one letter for coding lists of nonterminals $M[1]$, $M[2]$,..., and terminals $m[1]$, $m[2]$, etc.

This formal description of chain interactions in terms of T-FIN, apart from the needs of multicast systems, could yield one other useful result. It shows a way to provide formal grammars with synchronizing features and a notion of time.

5.3 Virtual Clothing

An interactive clothing system is one of the basic components of virtual reality. The clothing is implemented via a design that allows for the creation of two-dimensional (2D) garment panels that can be put on an actor in a realistic way. We show how to ease this time-consuming task by utilizing a hybrid of FIN with cellular automata (CA). We base our approach on the fact that the draping fabric shows a feature of natural parallelism. On a very short time scale, behavior of any node point of a fabric depends only on the behavior of its closest neighboring node points. Therefore, distant regions of the fabric can be simulated independently. Based on the problem's natural parallelism, we can design a special 2D FIN of uniform, locally connected, hybrid automata that execute in parallel a particle model of draping behavior of clothes on a standing mannequin (Tarakanov and Adamatzky, 2000).

5.3.1 Problem Description

Traditionally, clothes are made in two steps: 2D garment panels are cut out of the fabric and a mannequin or a human model is used to evaluate the results of the design. A similar procedure is employed in computer animation of clothes on artificial actors: 2D panels are designed and joined together, and then the garment is put on an actor and the movement of the clothes is simulated via animation. Fitting clothes on the actor and animating them both require simulation of the physical properties of the fabric (e.g., see Frisken-Gibson, 1999; Volino and Thalmann, 1998).

One problem of virtual clothing is that given 2D patterns of a garment and a human body, one must predict and visualize the 3D shape of the clothes according to the form and posture of the body.

There are advanced tools for virtual clothing that allow for interactive design of garments and thus substantially increase efficiency. These computer-based techniques are based on triangulation of imitated clothes, use of geometric constraints and application of finite difference and finite element methods (see, e.g., Volino and Thalmann, 1998) or on particle-based methods (see, e.g., Witkin, 1997). More often, geometry-based techniques are combined with a physics-based approach to utilize the realistic accuracy of physical techniques and to reduce the computation time to that of geometrical methods. These tools are actively used in the textile and fashion industries, computer animation, and virtual reality, and can certainly be utilized in Internet commercial applications, e.g., virtual clothing online shops.

In the particle-based approach, a garment is discretized into point-like masses, and mechanical forces are represented via local interaction between the particles. A fabric is seen therefore as a layer cloud of particles locally connected by elastic fibers, a so-called mass-spring structure.

In this section we develop this particle-based approach a bit further and utilize the inherent specifics of the problem. We make an assumption, quite correct from the physical point of view, that coordinates of every node point X of the fabric in 3D Euclidean space are changed depending mostly on coordinates of the node points connected to the point X. That is, the global behavior of the garment is described by local transformation rules applied to each node of the fabric in parallel.

As soon as the physical behavior of the simulated system is local and parallel, we can represent it using the concept of programmable matter (see, e.g., Toffoli, 1998). This concept, roughly speaking, states that the architecture of a computing device that solves a problem must conform to the architecture of the problem itself. CA is the first candidate for the parallel simulator. CA is already effectively applied to problems of image processing and computational geometry (see, e.g., Toffoli and Margolus, 1991), in particular, and analysis of complex systems (see, e.g., Kuznetsov, Milayev and Tarakanov, 1999, Weimar, 1998), in general. To exploit the full power of CA we should make all states of a simulated system discrete. However, it would not be particularly correct to get rid of the continuous coordinates of garment node points. Therefore, we choose a compromise between the accuracy of the continuous and the efficiency of the discrete.

We assume that the system consists of a discrete number of elements, which evolve in a discrete time but take continuous states. Thus, the system is built of both discrete and continuous components, and belongs, therefore, to a class of hybrid systems. In the light of these assumptions, it would be more appropriate to consider a uniform network of hybrid automata whose states are real-valued variables (Johansson et al., 1999). Networks of continuous-state automata are applied to problems from very different fields of science: from inversion of Voronoi diagrams (Adamatzky, 1994) and image processing (Yang T. and Yang L.-B., 1997) to simulation of etching processes (Zhu, 2000) and earthquakes (Hainzl et al., 1999). These CA networks have been developed further as adjuncts of cellular neural networks (Chua and Yang, 1988) and cellular immune networks (Tarakanov and Dasgupta, 2000).

Section 5.3 is organized as follows. In Section 5.3.2 we design a hybrid automata model for virtual clothing. A formal model is constructed in Section 5.3.3. The advantages of the model are considered in Section 5.3.4. Some drawbacks and potentials of the model and ideas for further development are outlined in Section 5.3.5.

5.3.2 An IC Scheme of Virtual Clothing

Virtual clothing aims to find a best possible correspondence between the planar geometry of a garment design and the nonplanar geometry of a body that wears the final product. We solve the problem using a modified FIN, a network of locally connected uniform CA with external inputs and real-valued states. In the design of the model we use the fact that cloth draping has a feature of natural parallelism. At a very short time scale, behavior of any node point of a fabric depends only on the behavior of its closest neighboring node points. Therefore, distant regions of the fabric can be analyzed, or simulated, independently and in parallel.

First of all, we represent the garment by a system of locally connected particles. Namely, every node point of a fabric is represented by a particle, ideally via a one-to-one correspondence. However, to speed up the simulation, one can map several neighboring nodes of the fabric to one particle of the model system. Then we map the particle system to a 2D automaton mesh. This representation is shown in Figure 5.5. The automata corresponding to garment particles and stitches are shown by different gradations of gray. Automata representing stitches can "glue" to, or rather be sewed to, edge automata of the same or another panel.

Every automaton corresponding to an inner particle of a garment panel has four neighbors. Edge automata have one, two, or three neighbors each. A body to which the garment is fitted is described by a mesh of triangles, as is traditional for 3D graphics applications. We also assume that the motion of fabric panels is composed of three stages: rigid displacement, free fall and swing, and local elastic deformations, as shown in Figure 5.6.

The coordinates of a particle, represented by an automaton x, are updated depending on the coordinates of particles, represented in states of closest neighbors of the automaton x. This deterministic local update is an essential feature of IC-based virtual clothing. Every automaton updates the coordinates of its own particle depending on the coordinates of the particles stored in the neighboring automata, the external forces, the forces between neighboring particles, and the geometries of both garment and mannequin's body:

1. If a particle collides with any of the body's triangles, then the particle is assumed to be fixed, and its coordinates are not updated.
2. If a particle has to be sewed with another particle, then the particle moves in rectilinear motion toward that particle.
3. A particle executes rectilinear motion under a gravitational force (Figure 5.5a) if all the neighboring particles execute such motion.
4. A particle swings and moves along an arc centered at its suspension point (Figure 5.5b) if each of its neighbors is either fixed or swinging.
5. If a particle is not fixed, it also moves toward the more distant of its closest neighbors at some fixed distance Δh (Figure 5.5c); this last step simulates the elasticity of the fabric.

5.3 Virtual Clothing   103

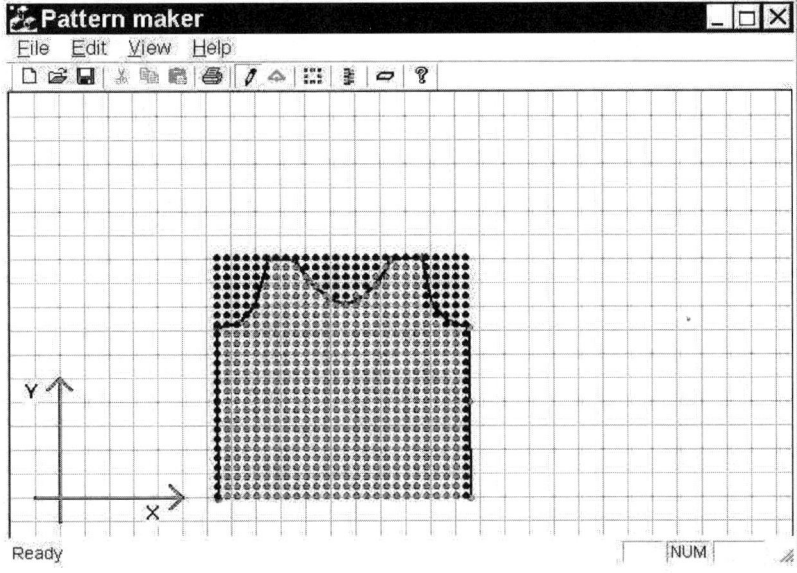

(a)

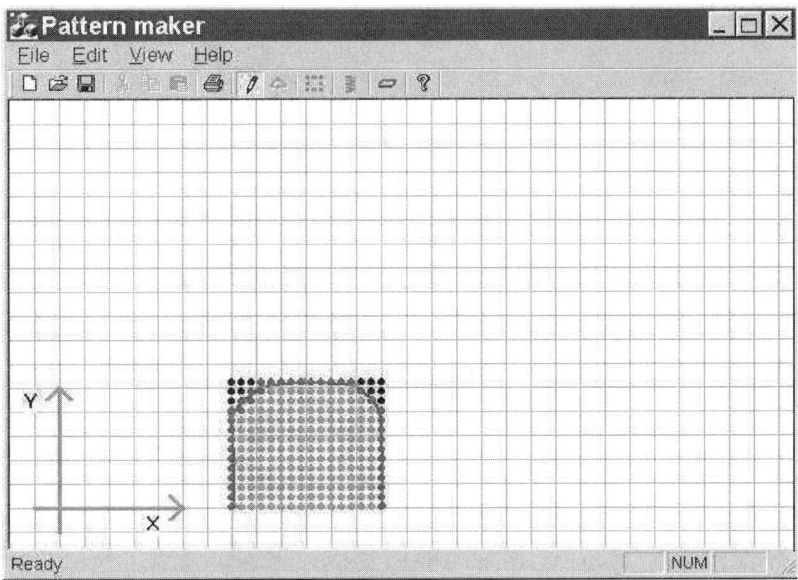

(b)

Figure 5.5 Representation of garment patterns as a system of connected particles: (a) Front and back panels of a T-shirt; (b) Panels of the left and right sleeves.

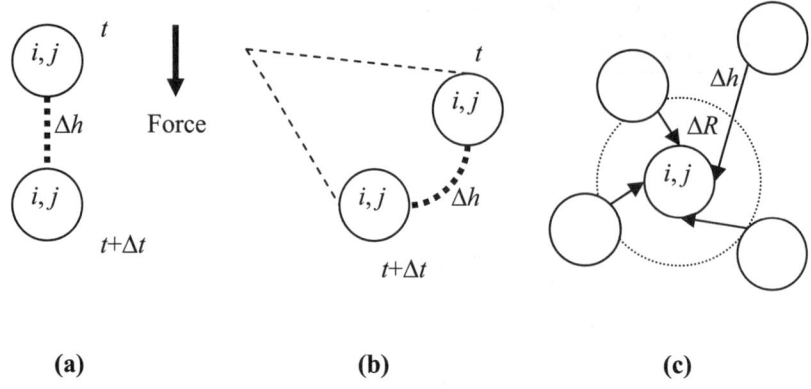

Figure 5.6 Types of particle motion: (a) free fall; (b) swing; (c) elastic motion.

5.3.3 Formalization

Now we are ready to give a formal definition of an automata network, or a hybrid FIN for virtual clothing. A hybrid cellular automaton, or hybrid FIN, is a set

$$\langle L, E, S, C, I, u, f, g, \mu \rangle,$$

where L is an integer lattice of uniform automata, connected locally according to a neighborhood template u; every automaton has its own external input (collision detection) $z \in I$, with the state set $I = \{$ False, True$\}$; E, S, and C are sets of automaton states, or components of the automaton's compound state; and f, g, and μ are automaton state transition functions.

We use a cruciform 4-automaton neighborhood as shown below:

$$u(x) = \{ y \in L : |x-y| = 1 \},$$

where $|\ldots|$ is a distance in a metric L_1.

All automata of the lattice update their states simultaneously in discrete time. Every automaton takes its state from the direct product of the sets E, S, and C:

$$x^t \in E \times S \times C,$$

where

$$E = \{\,Hole,\ Sewing,\ Drop,\ Swing,\ Fixed,\ Collided\,\},$$
$$S = \{\,non\_Sewed,\ Sewed\_1,\ Sewed\_2\,\},$$

are discrete, while C is continuous.

Elements of automaton state sets E and S represent various physical states of a garment's node points, or particles:

1. *Hole*: no particle of a garment;
2. *Sewing*: a "sewing" particle of a garment, which has to be sewed to another particle, according to stitch information;
3. *Drop*: a particle in a free fall;
4. *Swing*: a swinging particle;
5. *Fixed*: an immobilized particle;
6. *Collided*: a particle that has collided with the mannequin's body;
7. *non_Sewed*: a particle that has not yet been sewed;
8. *Sewed_1*: a particle that is sewed first and becomes a leading particle in a pair of two sewing particles;
9. *Sewed_2*: a particle that is sewed second and becomes driven by the first (*Sewed_1*) particle.

An automaton's compound state x^t at time step t can be seen as a set of three component states

$$x^t = \langle p^t, a^t, c^t \rangle,$$

where $p^t \in E$ and $a^t \in S$ are finite states of the automaton x, and $c^t \in C$ are coordinates of the particle, represented by the automaton x, in a three-dimensional Euclidean space $C \subset R^3$.

Further, we will write overall states of automaton x's neighbors at time step t as $u(x)^t$, and designate separate states of their components as $u(p)^t$, $u(a)^t$, and $u(c)^t$, respectively.

Every automaton x updates its states as follows:

The first component of the automaton state $x^t = \langle p^t, a^t, c^t \rangle$ is calculated by the function f,

$$p^{t+1} = f(p^t, u(p)^t, z^t),$$

where z^t is a state of external input of the automaton x:

$$p^{t+1} = \begin{cases} Hole \text{ if } p^t = Hole; \\ Sewing \text{ if } p^t = Sewing; \\ Fixed \text{ if } p^t \in \{Fixed, Collided\}; \\ Drop \text{ if } p^t = Drop \text{ and } \forall y \in u(x): p^t_y = Drop; \\ Swing \text{ if } p^t \in \{Drop, Swing\} \text{ and } \exists y \in u(x): p^t_y \in \{Fixed, Sewing\}; \\ Collided \text{ if } p^t \in \{Drop, Swing\} \text{ and } z^t = True. \end{cases}$$

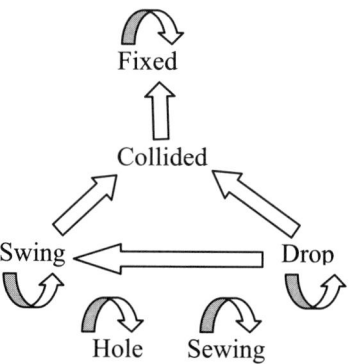

Figure 5.7 State transition diagram of a component p of an automaton compound state. The edges of the graph represent transitions $p^t \to p^{t+1}$.

A graph of automaton x state transition diagram is shown in Figure 5.7, which consists of three components: two of them are singletons, and the third includes four automaton states.

The second component a^t of the automaton state represents the current sewing status of the particle. This component is calculated by the function g:

$$a^{t+1} = g(a^t, y^t_{sew}, h_{sew}),$$

where h_{sew} is a real constant threshold of sewing, and automaton $y_{sew} \in u(x)$, is a neighbor of the automaton x such that the particle represented by x must be sewed to the particle represented by y_{sew}:

$$y_{sew}: p_x = p_{sew} = Sewing.$$

5.3 Virtual Clothing

The component a^t is updated by the following rules:

$$a^{t+1} = \begin{cases} Sewed\_1 & \text{if } a^t = non\_Sewed \text{ and } |c^t_{sew} - c^t| < h_{sew}; \\ Sewed\_2 & \text{if } a^t \neq Sewed\_1 \text{ and } a^t_{sew} = Sewed\_1; \\ a^t & \text{otherwise.} \end{cases}$$

This is a slightly simplified representation of the sewing of different particles. In reality it looks rather similar to the "antigene–antibody" paradigm of IC (Tarakanov and Dasgupta, 2000).

Updating of the third component c^t,

$$c^{t+1} = \mu(c^t, u(c)^t, p^t, a^t),$$

reflects the motion of the virtual particle, represented by the automaton x. The function μ is implemented in several steps. The particle coordinates are calculated as follows:

$$c^{t+1} = c^t + (v_1 + v_2)\Delta h,$$

where $v_1, v_2 \in R^3$ are directions of particle motion, and Δh is a real-valued step of the motion. The vector v_1 deals with the motion of a particle under the forces external to the garment:

$$v_1 = \begin{cases} 0 & \text{if } p^t = Hole \text{ or } p^t = Fixed; \\ v_{drop} & \text{if } p^t = Drop; \\ v_{swing} & \text{if } p^t = Swing; \\ v_{sew} & \text{if } p^t = Sewing \text{ and } a^t = non\_Sewed; \\ v_{sewed} & \text{if } p^t = Sewing \text{ and } a^t = Sewed\_1; \\ v(y_{sew}) & \text{if } p^t = Sewing \text{ and } a^t = Sewed\_2. \end{cases}$$

where

$$\begin{aligned} v_{drop} &= (0, -1, 0), \\ v_{sewed} &\in \{v_{drop}, v_{swing}\}, \\ v(y_{sew}) &= v_{sewed}(y_{sew}), \\ v_{sew} &= c(y_{-sew}) - c^t / |c(y_{-sew}) - c^t|, \\ v_{swing} &= A(c^t - c^t_s) - |A(c^t - c^t_s) - c^t|, \end{aligned}$$

$A = A(Q, \alpha)$ is a matrix of rotation through angle α about the axis Q. The axis Q is calculated as a vector product

$$Q = (0, -1, 0) \times (c^t - c^t_s).$$

The angle α is computed as follows:

$$\alpha = \frac{\Delta h}{|c^t - c^t_s|},$$

where c^t_s is a suspension point of a particle, represented by the automaton x. This point is determined by the coordinates of swinging and fixed particles, represented by the automata from $u(x)$.

Intergarment forces are employed in calculation of the vector v_2:

$$v_2 = \begin{cases} 0 & \text{if } p^t \in \{Hole, Fixed\}, \\ v_m & \text{otherwise,} \end{cases}$$

where

$$v_m = [(c^t_m - c^t)/(|c^t_m - c^t|)],$$
$$c^t_m = \max_{y \in u(x)} |c^t_y - c^t|.$$

The external input value z^t of the automaton variable x indicates whether the particle, represented by the automaton x, coincides with the mannequin: $z^t = True$, or $z^t = False$. In our model, the values of automata external inputs are calculated using a partitioning of the mannequin's surface by a variety of bounding boxes (Liu et al., 1995). Namely, we represent a mannequin's surface by a triangular mesh and cover it by the general bounding box. Then we partition the bounding box into a set of equal-sized boxes, or cells, where every cell contains a list of the local triangles covered by it. So, at every step of the process of clothing the mannequin we determine which cells contain the segment ($c^{t+1} - c^t$) and sort only a few local triangles to detect a collision. In our current setup, the mannequin is motionless. Therefore, we do not need to update the partition during the simulation and can set up a correspondence between the bounding box cells and sublattices of the hybrid automata lattice.

The initial conditions of automata evolution are worth discussing. First, two components of the automaton state are obviously set to the following states:

$$p^0 \in \{ Drop, Hole, Sewing \},$$
$$a^0 = non\_Sewed.$$

Several possibilities exist for determining the initial conditions of the third component c^0 of the automaton state. One can simply put all planes horizontally above a mannequin and let them fall free. However, we can reduce computation time if we help the automaton and place panels over certain parts of the mannequin, rotate the panels or even bend them into a torus, as we do, for instance, with sleeves and a skirt (see Figure 5.8a).

5.3.4 Numerical Results

To verify our theoretical ideas on the IC approach to virtual clothing, we designed a software implementation of the model. The software system is built of two modules: an interactive graphic interface for 2D design of garment panels and a module of dynamical draping.

The simulated planes of the fabric consist of 2420 node points, the front and back pieces of the T-shirt are 25×26 nodes each, the left and right sleeves 15×16 nodes each, the front and back pieces of the skirt are 20×16 nodes each. In the experiments we fit clothes on a female mannequin with the following parameters: height 172 cm, bust 85 cm, waist 62 cm, and hips 90 cm.

The mannequin is in its standing position during the entire period of simulation; i.e., no movements of the mannequin are involved in the simulation.

Initially we cut out a T-shirt, built of front and rear panels (Figure 5.5a), and two sleeves (Figure 5.5b). The sewing points are specified by a special procedure according to a list of stitches. The result of the simulation is shown in Figure 5.8, where you can also see a skirt, consisting of front and back panels.

The results of the simulation are encouraging. As in situations with CA approaches for modeling reaction–diffusion and excitation media and experiments with lattice gas (e.g., see Adamatzky, 1994, 1997), our current IC approach has proved to be effective. The problem of making a virtual T-shirt and skirt is solved by our automata network model in seven seconds on a standard computer with CPU of 400 MHz and 64 MByte memory (actually, all executables and data files are fit into 1.5 MByte). This yield gives an almost threefold increase in speed over traditional techniques that employ a triangular mesh representation of fabric.

Involvement of more realistic panels of fabric and the physics of garment behavior, particularly incorporating weight and self-collisions between garment particles as well as more realistic interaction with the mannequin, will certainly slow down the computation. However, this automata-based approach leaves open several ways to improve the performance of the technique. We can speculate that via optimization of mathematical representation and employing advanced technology, like MMX, XMM, or DSP (digital signal processor) we can simulate dressing of a realistic body in real samples of the garment in quite a reasonable time, possibly less than 10 seconds. Moreover, hardware implementation of the IC opens a promising way to perform realistic virtual clothing online.

110 Modeling of Natural and Technical Systems

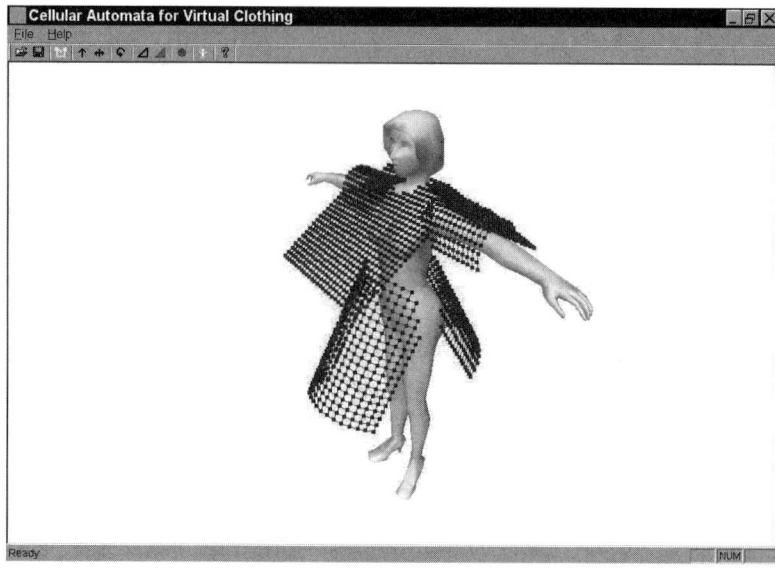

(a)

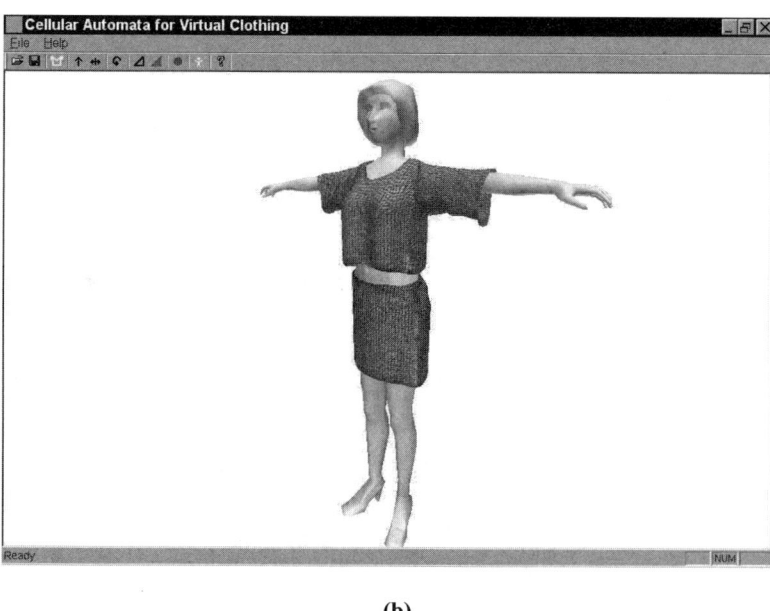

(b)

Figure 5.8 An interface of an IC-based system of virtual clothing. (a) Initial configuration of 2D panels of the T-shirt and the skirt in 3D space. (b) Resultant virtual T-shirt and skirt.

5.3.5 Discussion

To speed up the computational process in virtual clothing we undertook two subsequent transformations, which also evoked some shifts in paradigmatic contexts. At the first stage we represented a piece of fabric as a mesh of particle-like nodes, where each node is linked to its four closest neighbors by elastic fibers. We also assumed that the particles of the mesh execute a few basic motions when a mannequin is draped by the fabric: The particles either have a free fall, or swing, or remain motionless; sewing and elastic movements are also taken into account. As soon as the behavior of each particle is determined by the local structure of the mannequin's surface and the local behavior of the space occupied by the particle's closest neighbors, the time dynamic of the mesh can be reasonably simulated by a computing device with massive parallel architecture. So, at the second stage of problem transformation, we represented the particle mesh in a homogeneous, locally connected network of uniform automata. Each automaton of the network is a hybrid one, because it updates its states in a discrete time, but some components of its compound state take continuous values. This automata model, even implemented in standard software and executed on a serial computer, showed a drastic decrease in simulation time and reduction of memory requirements, hence preserving accuracy.

The IC approach has potential not only for the garment industry and for virtual reality, but also in the development of "smart clothes."

Smart clothes, by analogy to smart matter (Hogg and Huberman, 1998), feature fast local responses to changing conditions of environment and decentralized control, and may exhibit some kind of global knowledge emerging in the collective dynamic of cloth elements. For sets of particular parameters of interactions between cloth elements, the cloth itself is able to produce intelligent autonomous reactions and thus protect a human wearing it and enhance his/her physical capabilities via self-organization of garment structure (Maselko,1996).

Undoubtedly, the lion's share of the problem of virtual clothing is beyond the considerations of the present chapter. Some of these problems are the sliding of garment particles along the body; special features of the garment that arise as a result of intergarment collisions and/or collisions with the mannequin, e.g., formation of fine folds and wrinkles of the garment; garment animation during mannequin movements; deformable clothes; models of clothes made of different types of fabric with various physical characteristics, e.g., density. All these problems may be considered further.

Also, detection of collisions between fabric particles and the mannequin surface was implemented by a computing device external to hybrid FIN. This may present a potential drawback of the model. However, we see an elegant solution in representing a mannequin also as a hybrid FIN; this idea is similar to a linked volumes model (Frisken-Gibson, 1999) and so-called surface CA (Gobran and Chiba, 1999).

Nevertheless, the proposed IC approach, even in a nutshell, allows a fast and accurate solution of the virtual clothing problem. Therefore, further development will also deal with fine-tuning of the original technique.

6
Applications

6.1 Detecting Dangerous Situations in Near-Earth Space

Near-Earth space is being increasingly contaminated by hundreds of useless artificial objects. This "space junk" is a serious potential hazard to future space missions. Thus, accurate hazard assessment of near-Earth space is crucial for mission safety.

However, the assessment problem has not yet been solved on a global scale, primarily because of the following three difficulties:

- the complicated rules of orbital motion, especially the relative motion of satellites;
- high-order motion dynamics;
- the representation of 3D motion space on 2D displays.

These difficulties can be overcome by the IC approach, since it can perform analysis and visualization of global ballistic situations in near-Earth space. This approach uses our results on the theory of the orbital hodograph, which reduces the dynamics of orbital motion of satellites to the relationship between geometric invariants in specially constructed planes (Tarakanov, 1986, 1992). Such planes can also be treated as shape spaces of the IC to model the parameters of orbital hodographs.

Applications of this approach, together with the theory of the orbital hodograph, facilitate the successful resolution of the above-mentioned difficulties by the IC. Thus, the approach provides the user with the essence of the assessment in comprehensive terms. Appropriate IC technologies could also be made capable of providing real-time decision-making in autonomous planning and control of space situations.

Consider, for example, the following task. Over the set of all satellites and space junk, detect only those pairs whose orbits intersect.

To solve this task, represent every orbit as a one-link FP (monopeptide) with quaternion unit vector (Q-vector)

$$[Q]^T = [-\sin\omega, \; c_1\cos\omega + s_1, \; s_1\cos\omega - c_1, \; 0],$$

where ω is the perigee argument (angle) of the orbit, and c_1, s_1 are constants determined by the fixed valence angle η of the FP (see Section 2.1.1).

Let every FP have an *active center* with an *active center matrix A* and some *environmental matrix S* that is the same for the all FPs, where p is the focal parameter and e is the eccentricity of the orbit:

$$A = \begin{bmatrix} -e & 0 & 0 & 0 \\ 0 & c_1 e & s_1 e & 0 \\ 0 & s_1 p & -c_1 p & 0 \\ 0 & s_1 \dfrac{1-e^2}{p} & -c_1 \dfrac{1-e^2}{p} & 0 \end{bmatrix},$$

$$S = -\begin{bmatrix} 2 & 0 & 0 & 0 \\ 0 & 2 & 0 & 0 \\ 0 & 0 & 0 & 1 \\ 0 & 0 & 1 & 0 \end{bmatrix}.$$

Let the binding energy between two FPs Q_1, Q_2 be computed as

$$w(Q_1, Q_2) = -[Q_1]^T A_1^T S A_2 [Q_2],$$

where A_1, A_2 are the corresponding active center matrices. Then, in accordance with the equation

$$[Q]^T A^T = \left[e\sin\omega, \; e\cos\omega, \; p, \; \dfrac{1-e^2}{p} \right],$$

we obtain the following result:

$$w(Q_1, Q_2) = \dfrac{p_1}{p_2}(1 - e_2^2) + \dfrac{p_2}{p_1}(1 - e_1^2) + 2e_1 e_2 \cos(\omega_2 - \omega_1).$$

Let the threshold of binding be $w_h = 2$. Then, according to (Tarakanov, 1986), bound FPs with $w \leq w_h$ correspond to the crossed orbits, where $w = w_h$ corresponds to the orbits that touch.

Given the possibility that a set of such FPs can interact and bind within the IC shape space, the results of this self-organization are complexes of FPs that select, from the set of all specified orbits, only the crossing and the touching pairs of orbits. This approach has been extended for the determination of spacecraft visibility from a point on Earth, as well as for the determination of the observability of particular areas on Earth's surface from a spacecraft.

A simulation of such an IC-inspired approach using a standard PC over a set of more than 100 specific orbits has showed that it allows the clear representation of the entire global ballistic situation. Its quick analysis results in the detection of dangerous objects as follows:

Table 6.1 shows a fragment of standard space navigation data from the year 1985, where "M" denotes the satellite "Molniya," i is the angle of the plane of orbit to the plane of equator, and Ω is the ascending longitude node.

Computed on the basis of these data, Figure 6.1a, b represents the fragments of a global space situation mapped onto the 2D shape spaces of the IC.

Satellite, Time of perigee	p (km)	e	$\omega°$	$i°$	$\Omega°$
M-3 #43, 23 03 39	13079	0.71261	279	64.7	261
M-3 #52, 01 53 59	12670	0.72339	282	62.8	205
M-1 #67, 23 28 45	14396	0.68317	273	63.8	158
M-1 #71, 12 23 22	14289	0.67967	278	63.3	122
M-1 #78, 15 03 11	13467	0.70229	280	63.0	136
Shuttle, 22 56 44	6698	0.00042	285	49.6	123
Imuse, 11 10 38	42166	0.00008	79	1.3	131
Fleetsat, 15 40 56	42167	0.00027	168	0.6	42
Ferret, 02 06 52	6708	0.00755	43	97.4	331
...	...	...	...	...	...

Table 6.1 Parameters of the orbits.

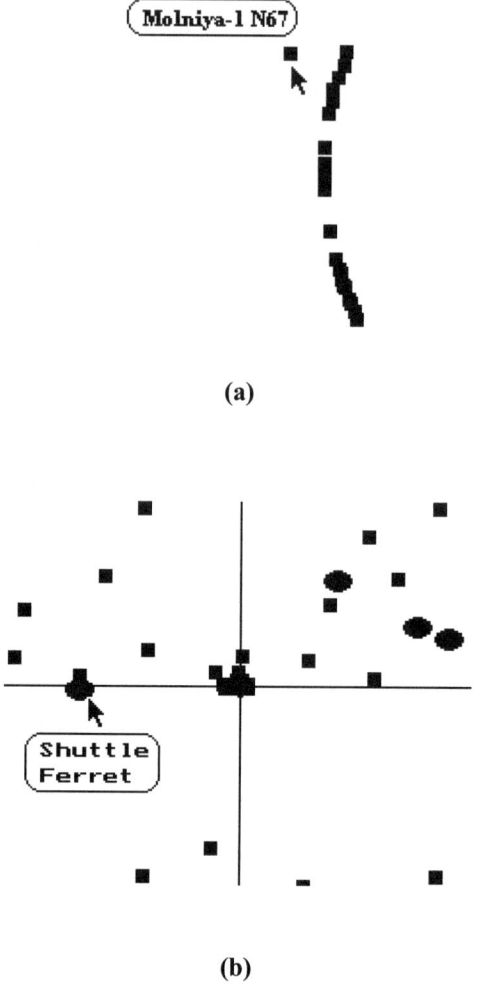

Figure 6.1 Global ballistic situation on shape spaces of IC

In Figure 6.1a, every orbit is represented by an FP (square). These FPs are able to bind according to their proximity in the shape space of the IC. Figure 6.1a clearly detects three groups of bound FPs corresponding to three systems of communication satellites: "Molniya-1, 2, 3." The IC also clearly detects that something is wrong with the satellite "Molniya-1 #67," because the corresponding FP is unable to bind with any other FP.

In Figure 6.1b every pair of intersecting orbits is represented by one square, and every pair of satellite positions on these orbits is represented by one circle. If an FP-square binds with an FP-circle, and this combination binds with the horizontal line, then the distance between two corresponding satellites is dangerous. Therefore, Figure 6.1b detects one dangerous distance, between "Shuttle" and "Ferret," among all possible pairs of satellites.

This IC approach is also well suited to the prognosis of dangerous situations, because, in contrast to the complicated rules of orbital motion, FP-circles in Figure 6.1b exhibit circular motion, while FP-squares do not change their positions at all.

This IC approach has also been used for the determination of visibility of several satellites from Earth, as well as for the observation by satellite of several areas on Earth's surface.

6.2 Complex Evaluation of Ecological and Medical Indicators

To provide policies of sustainable development, administrators need objective and complete information about their immediate ecological situations and changes thereof. Usually, such information is obtained by direct or indirect evaluation of the pressure on the environment from separate industrial factors: pollution of air, water, soil, etc. However, this rather close approach permits neither comparison of regions in general, nor standardization, automation, and computerization of the procedures of ecological evaluation.

The ecological atlas of St. Petersburg (Gorelick et al., 1992) is a pioneering example of a radical new approach to complex ecological evaluation (CEE). This atlas includes 10 special maps, specifically representing ten major environmental factors. In addition, the eleventh map of this atlas summarizes these factors, thus becoming a map of the general quality of the environment. This unification of rather heterogeneous factors has been made possible by at least two main rules:

1. Every factor is considered to be a constant within a small area of standard size;
2. Each measured factor is represented as a relative mark in some discrete scale.

Therefore, the essence of the CEE consists in obtaining some general ecological classification that integrates a large number of initially measured special factors. It is worth noting that a similar problem is of great importance to the World Health Organization (WHO), where the mark of general health status is also called an "indicator" or "index" (Catalogue, 1996).

The starting point for mathematical formulation of the CEE problem is the scheme of the proposed approach in Figure 6.2, where:

- $i=1,...,n$ is the number of a map (layer) of the special ecological mark (SEM);
- $x_1,...,x_n$ are values of the SEMs (usually displayed in the map as a color of the corresponding square);
- z is a complex ecological evaluation (CEE).

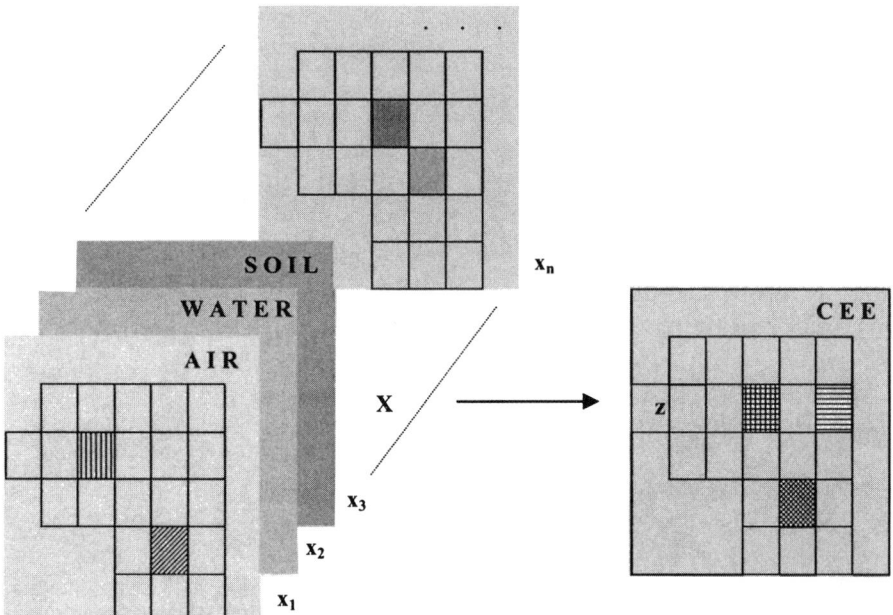

Figure 6.2 CEE scheme.

We also use the term *indicator* as a synonym for SEM, or x_i, and the term *index* as a synonym for CEE, or z. This separation is conditional, because the problem of determining an indicator by the given set of measured ecological factors (data of ecological and/or medical monitoring) is formulated and solved by the same scheme as the CEE problem.

Based on this scheme, the CEE problem has been stated and solved by (Kuznetsov, Miliaev, and Tarakanov, 1999) using an IC approach to pattern recognition (see Chapter 3). We consider several applications of this approach.

6.2.1 Ecological Atlas of the City of Saint Petersburg

The IC approach to CEE and the corresponding computer system based on it were first tested using the data from the ecological atlas of St. Petersburg

(Gorelick, Kuznetsov, and Khvorov, 1992). This atlas contains the CEE map, which has been elaborated on the basis of specially selected maps. These maps contain representations of contaminations of air and soil, as well as conditions of flora and ornithofauna. In turn, these special maps were elaborated on the basis of expert evaluation and by using definitely calculated coefficients. At that time this approach did not have a rigorous mathematical foundation.

In order to verify the obtained results, a learning sample was formed on the basis of these special maps. Variables of the learning sample, folded to the matrix $A=[a_{ij}]$ with dimension 2×2, are given in Table 6.2, where a_{11}, a_{12} are contaminations of air and soil, and a_{21}, a_{22} are conditions of flora and ornithofauna.

The results of CEE recognition are given in Table 6.3, where zero SEM (indicator) implies that the appropriate information was unavailable. The Euclidean distance d_k between the tested vector and the CEE class $c=1,2,3,4$ was computed on the shape space of IC, and the minimum distance was underlined for every test.

Kind of mark→	SEM	SEM	SEM	SEM	CEE
mark→	a_{11}	a_{12}	a_{21}	a_{22}	z
(scale)→	(1–7)	(1–4)	(1–6)	(1–5)	(1–4)
District ↓					
1.1 Finnish railroad station	1	1	1	1	1
1.2 Warsaw railroad station	1	1	1	1	1
1.3 Bol'shoi prospect of Vas. Island	2	2	4	2	1
1.4 Bol'shoi prospect of Petrog. Stor.	1	2	4	2	1
2.1 Piskaryovka	3	3	2	3	2
2.2 Novaya Derevnya	3	4	2	1	2
2.3 Ribatskoye	3	4	4	2	2
2.4 Hares' Island	2	3	4	2	2
3.1 Lakhta	5	4	2	3	3
3.2 Ozerki	4	4	4	3	3
3.3 Staraya Derevnya	3	4	4	2	3
3.4 Obukhovo	3	4	3	3	3
4.1 Krestovski Island	5	4	6	4	4
4.2 Lake Dolgoye	6	4	2	3	4
4.3 Airport Pulkovo	6	4	5	3	4
4.4 Kupchino	5	4	4	3	4

Table 6.2 Learning sample from ecological atlas of St. Petersburg.

Test #	a_{11}	a_{12}	a_{21}	a_{22}	d_1	d_2	d_3	d_4	CEE = c c: min d_c
1	6	4	2	3	6.2	5.2	4.2	<u>3.2</u>	4
2	5	4	5	3	2.6	1.9	2.0	<u>1.6</u>	4
3	4	4	4	3	2.4	1.0	<u>0.9</u>	1.7	3
4	5	4	5	2	3.7	<u>2.2</u>	2.8	2.8	2
5	3	4	4	2	3.0	<u>1.4</u>	2.3	3.4	2
6	3	4	3	1	3.8	<u>1.8</u>	2.6	3.7	2
7	1	2	4	2	<u>1.8</u>	2.8	3.5	4.0	1
8	5	4	0	4	8.1	7.7	6.8	<u>6.2</u>	4
9	3	4	4	0	5.3	<u>4.1</u>	5.0	5.8	2
10	6	0	0	3	10.5	10.8	9.8	<u>8.5</u>	4
11	5	4	0	0	5.8	5.6	4.9	<u>4.3</u>	4
12	4	4	0	0	5.1	4.5	<u>3.8</u>	4.4	3

Table 6.3 Results of recognition.

It is interesting to note that experts have assigned different CEE marks (indices) to the same set of indicators $\widetilde{X}$ = [3 4 4 2] in the learning sample in Table 6.2: mark 2 to the district 2.3 and mark 3 to the district 3.3. However, the system recognized this set as having index 2 in Table 6.3, test 5. Remarkably, this CEE is more consistent with common sense. This was confirmed by comparative analysis of indicators a_{22} from Table 6.2, as well as inspection of the district. Note that the system recognized the CEE successfully even in the absence of the half of the data (see Table 6.3, tests 10–12).

Therefore, the ecological atlas of St. Petersburg demonstrates the practical applicability of the proposed IC approach to the CEE . Moreover, the model is able to process incomplete, imprecise, and even conflicting data.

6.2.2 The Ecological Atlas of the City of Kaliningrad

The next application of the model was the computation of the CEE map of Kaliningrad, based on data obtained by ecological monitoring (Kuznetsov, Gubanov, Kuznetsov, Tarakanov, and Tchertov, 1999).

In this case, all work from the very beginning was dedicated to the goal of developing the CEE and its map. Special maps were developed as subject maps of particular factors of pressure on the environment or a particular component of response to these pressures. All these indicators were represented in relative scales.

The whole city was partitioned into a 17×23 squares grid with 1 km/side by latitude (L) and longitude (B). Eight maps were involved in the processing: (1) the map of geomorphology and landscape; (2)–(4) maps of contamination of air, water, and soil; (5) the map of acoustic conditions; (6) the map of electromagnetic fields; (7) and (8) maps of the conditions of flora and ornithofauna.

A map of radiation was not considered, because radiation in the city is low and had no significant effect on the environment.

According to the IC approach to supervised learning (see Section 3.2.2), the CEE map has been computed on the basis of the learning sample, which has been given by experts (Table 6.4, where squares of the learning sample are framed).

As a result, the computed CEE map has been highly appreciated by experts and has been included in the ecological atlas of the city of Kaliningrad (Kuznetsov, Gubanov, Kuznetsov, Tarakanov, and Tchertov, 1999).

	1	2	3	4	5	6	7	8	9	10	11	12	13	14	15	16	17	18	19	20	21	22	23
1	0	0	0	0	0	1	1	1	1	1	1	2	1	0	0	0	0	0	0	0	0	0	0
2	0	0	0	0	1	1	1	2	1	1	1	2	2	2	2	3	4	4	0	0	0	0	0
3	0	0	1	1	1	1	1	2	2	1	1	2	2	1	1	1	2	2	2	0	0	0	0
4	1	1	1	1	1	2	3	2	2	3	2	4	4	3	2	2	3	2	2	2	0	0	0
5	1	1	1	1	2	1	1	1	1	4	5	4	5	4	3	2	3	3	3	3	0	0	0
6	0	0	2	2	3	2	2	2	2	4	4	5	4	4	4	4	2	1	2	1	1	0	0
7	0	0	1	1	1	3	4	3	2	3	4	5	5	4	5	3	2	1	2	1	1	0	0
8	0	0	1	1	1	1	3	2	3	4	5	5	3	4	6	4	4	4	3	1	1	0	0
9	0	1	1	1	2	4	4	4	4	3	5	4	5	6	5	5	4	5	4	3	2	1	1
10	1	1	1	4	2	3	3	3	4	4	4	5	5	6	5	5	5	5	4	4	3	3	0
11	1	1	4	4	4	3	3	2	3	4	4	3	4	6	5	6	5	4	4	3	1	0	0
12	0	0	0	0	0	0	0	0	3	4	3	4	3	3	3	5	4	3	4	2	1	2	1
13	0	0	1	1	0	0	2	3	3	4	3	3	2	3	4	3	3	2	3	2	1	1	1
14	0	0	1	1	0	2	3	3	3	3	4	3	2	3	3	2	3	3	2	2	1	1	0
15	0	1	1	2	3	3	3	3	0	0	0	0	0	0	0	2	3	1	1	1	1	0	0
16	1	1	1	1	3	0	0	0	0	0	0	0	0	0	0	2	3	1	1	1	0	0	0
17	1	1	1	0	0	0	0	0	0	0	0	0	0	0	0	0	0	0	0	0	0	0	0

Table 6.4 CEE map of Kaliningrad.

6.2.3 Quality of Environment and Morbidity of Children for the City of Tula

The possibilities of the IC approach were clearly demonstrated in the ecological monitoring of Tula, the results of which are represented in Table 6.5. These results include values of three indicators (air, soil, water) for nine districts of the city. Table 6.6 contains the initial data from the medical monitoring of the same districts of Tula. Figure 6.3 represents all these data in the shape spaces of IC: $\{w1, w2, w3\}$, formed by unsupervised learning (see Section 3.2.3) separately for ecological (Figure 6.3a) and medical (Figure 6.3b) data.

District → ↓ Kind of SEM	1	2	3	4	5	6	7	8	9
Air	19	12	5	10	12	10	10	10	8
Soil	76	17	20	121	90	81	56	42	56
Water	51	10	10	37	51	38	<u>64</u>	64	51

Table 6.5 SEM for districts of Tula.

District → ↓Disease	1	2	3	4	5	6	7	8	9
Infections	5.7	4.8	4.6	4.4	<u>9.3</u>	7.2	<u>8.0</u>	3.6	6.3
New growth	0.1	0.2	0.0	0.2	0.0	0.3	0.1	0.1	0.1
Endocrine	0.3	0.7	0.3	1.5	0.6	0.8	0.4	0.5	0.7
Blood	0.4	0.2	1.8	0.6	0.6	0.6	0.2	0.1	0.3
Anemia	0.3	0.2	1.2	0.4	0.4	0.6	0.1	0.1	0.3
Nervous system	<u>12.8</u>	7.9	6.7	9.8	6.4	8.2	7.6	8.5	8.6
Blood circulation	0.3	0.3	0.1	0.5	0.5	0.7	0.1	0.3	0.2
Respiration	67.3	73.7	73.2	72.8	74.4	69.0	59.7	73.3	75.3
Asthma	0.3	0.9	0.2	0.3	0.3	0.4	0.9	0.3	0.5
Gastric-intestinal	4.7	2.7	1.5	2.5	2.9	5.5	<u>11.0</u>	4.8	2.1
Urinogenital	<u>1.3</u>	0.9	1.1	1.0	1.1	1.2	0.7	1.1	0.8
Skin	2.2	3.1	2.5	1.5	1.1	2.5	1.9	1.9	1.4
Bones and muscles	1.6	1.0	4.2	2.4	0.6	1.3	0.9	2.2	1.2
Inborn anomaly	0.3	0.6	0.6	0.5	0.6	0.5	0.1	0.5	0.7

Table 6.6 Morbidity of children for districts of Tula.

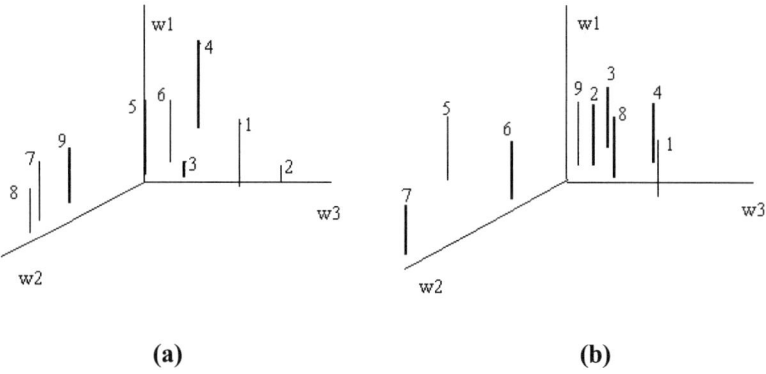

Figure 6.3 Tula districts on 3D shape spaces of IC: (a) CEE and (b) morbidity of children.

In Figure 6.3, the relative position of every point (top of the line) in the shape space is treated as the complex evaluation of the corresponding district relative to all other districts. For example, district 7 is chosen both in Figure 6.3a and in Figure 6.3b. But this district is characterized both by a bad quality of water (SEM underlined in Table 6.5) and by an increased rate of gastrointestinal diseases and infections (underlined in Table 6.6). Districts 1 and 5 are also selected in Figure 6.3b as having the extreme values of the indicator w3. This is how the IC approach focused attention of the experts on the relatively extreme rates of infections for district 5, as well as nervous and urogenital diseases for district 1 (underlined in Table 6.6).

Thus the IC approach has helped to find a strong correlation between the quality of environment and morbidity of children for several districts of Tula.

6.2.4 Similarity in Dynamics of Infection Morbidities in Russia

The initial data for this task were indicators of 65 infectious diseases in Russia during 1996–1997 (Health, 1997). Consider a fragment of the data in Table 6.7, where numbers of cases of infection correspond to the following diseases:

1. Bacterial dysentery;
2. Bacterial dysentery confirmed by bacteriological reports;
3. Bacterial dysentery caused by *Shigella Zonne*;
4. Tuberculosis (first detected);
5. Tuberculosis of respiratory organs;
6. Virus of hepatitis C;
7. Acute intestinal infections;
8. Scarlet fever.

124 Applications

In the shape space $\{w2, w3\}$ these diseases have coordinates, shown in Table 6.8.

The representation of data in Table 6.8 in Figle 6.4 allows for detection of obvious groups of diseases with the following numbers: {1,2,3}, {4,5,6}, {7,8}.

№ of infection → During ↓	1	2	3	4	5	6	7	8
1997 Jan–Apr (total)	14.5	9.4	2.6	16.0	15.1	17.1	19.1	13.9
1997 Jan–Apr (children up to 14 years old)	35.2	24.4	8.5	3.2	2.6	1.2	68.6	62.6
1996 Jan–Apr (total)	23.6	16.6	7.4	15.7	14.8	12.8	23.7	14.7
1996 Jan–Apr (children up to 14 years old)	60.4	45.4	25.1	2.9	2.4	1.9	85.1	65.3
1997 Jan–Sep (total)	41.3	27.0	9.3	37.8	35.4	37.3	47.6	24.7
1997 Jan–Sep (children up to 14 years old)	101	71.6	30.8	8.1	6.5	2.2	159	110
1996 Jan–Sep (total)	64.4	44.8	19.8	36.4	34.2	28.9	52.6	27.0
1996 Jan–Sep (children up to 14 years old)	163	122	65.9	7.2	5.8	1.7	174	120
1997 Jan–Nov (total)	53.5	36.1	14.7	47.3	44.3	47.1	57.4	31.7
1997 Jan–Nov (children up to 14 years old)	135	98.8	48.9	10.4	8.3	2.8	191	142
1996 Jan–Nov (total)	78.7	55.2	25.2	45.5	42.7	36.8	60.9	34.7
1996 Jan–Nov (children up to 14 years old)	201	152	84.4	9.1	7.3	2.2	201	155

Table 6.7 Infection morbidity rate in Russia during 1996–1997 (per 100 000 of population).

№ of infection → ↓ Coordinate	1	2	3	4	5	6	7	8
w2	0.13	0.05	−0.03	0.57	0.54	0.54	−0.12	−0.24
w3	0.48	0.43	0.37	−0.14	−0.12	−0.15	−0.50	−0.38

Table 6.8 Shape space coordinates of diseases.

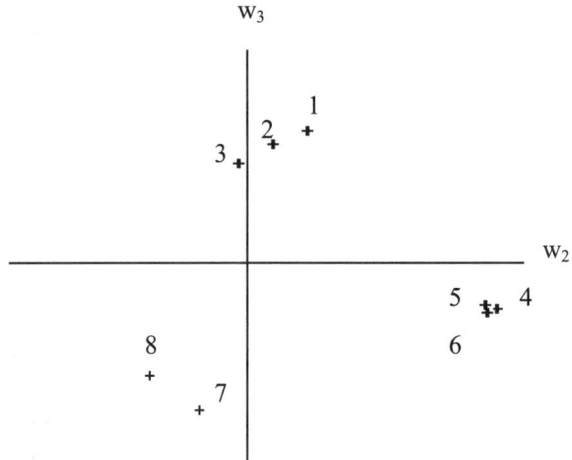

Figure 6.4 Indicators of the diseases on the shape space.

Figure 6.5 shows histograms of the corresponding diseases, drawn from the initial data of Table 6.7. These histograms confirm the obvious similarity of dynamics of infection morbidity within each of the three groups {1, 2, 3}, {4, 5, 6}, {7, 8}, and also confirm the obvious difference between diseases of the different groups.

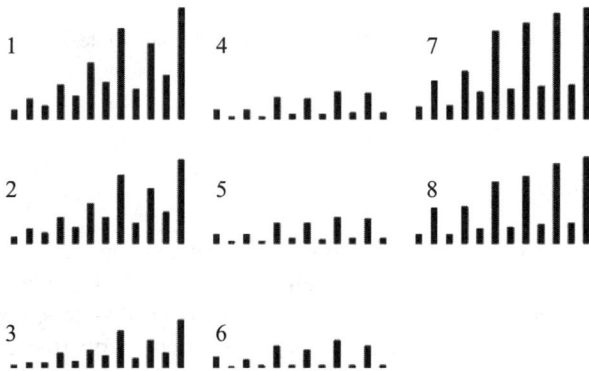

Figure 6.5 Histograms of the dynamics of morbidity rates.

126 Applications

Another confirmation of the similarities that have been detected is given by correlation coefficients between every pair of detected diseases in Table 6.9. In this table, correlation coefficients are enclosed in boxes when they are numerically close to 1.0 and correspond to the detected groups of diseases.

However, it is worth noting that the correlation coefficients do not allow us to represent groups of diseases as clearly as in Figure 6.4. Moreover, Table 6.9 also gives correlation coefficients close to 1 for all pairs between groups $\{1, 2, 3\}$ and $\{7, 8\}$, but Figure 6.5 shows the obvious difference in dynamics between groups $\{1, 2, 3\}$ and $\{7, 8\}$. Hence, correlation coefficients could be a means of confirming the similarities that have been already detected, but not a means of detecting similarities.

By means of the IC approach we have detected three groups of diseases with clearly similar dynamics. Moreover, the similarity between diseases $\{1, 2, 3\}$ (dysentery), as well as between $\{4, 5\}$ (tuberculosis) seems natural. However, the detected similarity between diseases $\{4, 5\}$ and $\{6\}$ (tuberculosis and hepatitis), as well as between diseases $\{7\}$ and $\{8\}$ (sharp intestinal infections and scarlet fever) seem unexpected and in need of the attention of specialists.

↓ infections →	2	3	4	5	6	7	8
1	1.00	0.99	−0.30	−0.33	−0.43	0.93	0.91
2		0.99	−0.33	−0.35	−0.45	0.93	0.91
3			−0.35	−0.38	−0.47	0.91	0.89
4				1.00	0.98	−0.49	−0.58
5					0.98	−0.52	−0.60
6						−0.60	−0.68
7							0.99

Table 6.9 Correlation coefficients between morbidity rates.

6.2.5 Conclusions

The results presented in Section 6.2 allow us to formulate and solve complex problems of ecology with the necessary mathematical rigor. This evaluation will become more complete and objective if apart from ecological factors, it includes indicators of their influence on health of the population. Furthermore, the IC approach allows us to formulate and to solve some complex health-related problems in a similar fashion (see, e.g., Tarakanov and Tumanov, 1998).

Another direction for further development of the IC approach for complex ecological evaluation is in computer-aided technology. In analogy to GIS technology, we propose to call this CEE system the ecoinformation system. Such systems, intended for computer-aided decision-making in the field of environmental protection, would be useful for administrators of different levels, as well as for professional ecologists.

6.3 Surveillance of the Plague in Central Asia

Natural plague foci in the former Soviet state of Kazakhstan cover an area of 130 million hectares, and over the past 50 years, they have been considered the most active plague foci in the world. Today the antiplague service in Kazakhstan is working under severe financial and organizational difficulties. Even against the background of general deterioration in medical services for the population, the epidemiological surveillance in plague foci has become irregular, the prophylactic antiplague measures, which used to ensure control of the situation and prevent risk of mass infection for man, have been significantly reduced. Recently, local and foreign workers have increased the level of human activity in the natural foci regions, often in connection with an intensive exploitation of natural resources (e.g., gas and oil). These activities are often organized by multinational companies, which increase the probability of plague cases being exported abroad.

Plague foci in Kazakhstan covering vast territories are characterized by different regulation mechanisms at the population, species, and community (biocenotic) levels, and they have not been studied systematically. The plague epizootic process is a complex, multicomponent, dynamic system. Both the behavior of particular subsystems of the plague epizootic triad (agent–host–vector) and the entire epizootic process in foci (taking into consideration the complicated interrelations of the above subsystems) still require thorough investigation.

6.3.1 Plague Features

The state of the agent is characterized by the following differential and diagnostic properties (Aikimbayev, 1994):

Qualitative Properties: the morphology of colonies, the susceptibility to bacteriophage (Pokrovskaya's homologous, heterologous pseudotuberculous), glycerin fermentation, rhamnose fermentation, denitrification/nitrification, pesticinogenecity, the susceptibility to pesticine, the presence of VW-antigens, the need of growth factors.

Quantitative Properties: the dependence on calcium at 37°C, the presence of an antigen of fraction 1 in the reaction of passive hemaglutination and immunoglobulin plague erytrocytic diagnosticum, growth on the medium with hemin, the integral property of virulence in white mice and guinea-pigs (LD-50, DCL, according to Kerber's calculation).

Most frequently, the state of the agent can be characterized only by its numbers expressed through indirect indices: the infestation of rodents, fleas, or samples (points) obtained not only from the given area but also from adjacent ones. The latter proves to be of great prognostic value when the spread of epizootic over the area is observed and a given section remains uninfected. The numbers of a microbe can be characterized indirectly by the number of animals in whose blood the plague antigen has been discovered. The facts of epizootic cyclicity may also be used in the process of forecasting.

The state of fleas—vectors of a plague microbe—is expressed through their numbers, their seasonal activity in attacking animals, through the sex and age composition of the imaginal phase, as well as with the help of the mass emergence of imago on animals and in openings of rodent burrows after hibernation, the ovipositor (its beginning, peak, and termination), the larvae hatch (its beginning, mass numbers, and termination) and other indices (Ageyev, 1975).

Other relative indices of numerousness include the index of abundance, the index of dominance, and the index of fidelity, the total numbers of fleas per hectare.

The state of hosts is characterized by the aggregate of factors of influence, the set of characteristics of processes taking place in populations at the biocenotic, population, and organism levels.

The biocenotic level of zoological monitoring suggests obtaining information from the following factors:

- species composition of hosts in a focus;
- degree of dominance according to the numbers and prevalence of species in the territory of a focus;
- relationship between the types of rodents, predators, and burrowing birds in a focus;
- level of interspecies contacts in the principal host burrows.

Zoological monitoring of the population level involves the following factors:

- population structure of the area of the principal host in a focus;
- spatial structure of colonies of the principal host in the territory of a focus;
- density of family burrows of the principal host per hectare, the correlation of the inhabited and uninhabited burrows and the fluctuation amplitude of these indices according to seasons and years, as well as the different phases of the population dynamics of animals;
- numbers of the principal host populations in different seasons and in terms of the density of the rodents population, as well as the amplitude to this index;
- cyclicity of the numbers of the principal host populations;

- age and sex structure of principal host population according to seasons, years, various phases of the population dynamics as well as in the nuclei and peripheral parts of colonies;
- reproduction of the principal host population;
- mortality rate of the principal host population;
- phenotypical structure of the principal host population in different parts of its area and its dynamics in different years and seasons against the background of the changes in the epizootic state in natural foci;
- structure of the principal host populations according to blood groups and its dynamics depending on plague epizootics;
- dynamics of the level of infection susceptibility of the principal host at various phases of the epizootic cycle of a focus;
- assessment of the physiological state of the principal host populations according to seasons, years, and at different phases of the population dynamic;
- correlation of mono- and polygamous families in the principal host population (the large gerbil).

The organism level of research involves obtaining the following information:

- determination of age, sex, and generative state of hosts;
- assessment of the physiological condition of animals of different ages and both sexes;
- level of individual susceptibility to infection in the principal host populations;
- mobility and the migration activity of animals of different ages and sexes in the principal host population.

The data characterizing the state of the members of the epizootic triad are classified according to the season (spring, summer, autumn), and sometimes (for example, in case of reproductive indices) according to months and even decades. The annual and seasonal growth indicating the speed and direction in the population changes are calculated from the basic seasonal indices of the members of the epizootic triad.

Weather, geographical, and space characteristics constitute the indices of *external factors*. The weather characteristics are the temperature of the air, soil, and precipitation; various hydrothermal coefficients (the relation of the amount of precipitation and the sum of temperatures over a certain period of time); the number of days below freezing without thaw; the sum of air temperatures over the period when it was above 10°C, 5°C; the recurrence of wind and the wind velocity according to compass points; the recurrence of the types of atmospheric circulation, the value of flood; etc.

The geophysical indices include the indices of geomagnetic activity, planetary and local, while the space characteristics include Wolf's numbers and the number of patches located along the Sun's central meridian.

6.3.2 AIS for Surveillance of the Plague

Complex interactions as are found in the plague system need new methods and approaches in order to improve the surveillance of epizootic and epidemic situations in the natural plague foci. Development of AIS has the potential to achieve this. The AIS is intended to improve risk analysis and underlying understanding of the space-time dynamics of the plague in Central Asia. Moreover, the AIS could be considered as a part of a general surveillance system for other infectious diseases that are reemerging through tourism and other international activities. Also the AIS could be considered as a model system for processing surveillance data on dangerous infections in all parts of the world.

Consider the architecture of the proposed AIS in Figure 6.6.

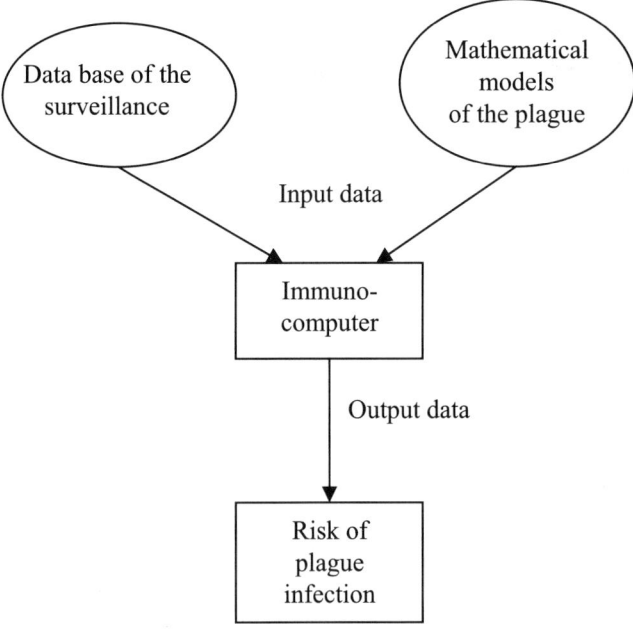

Figure 6.6 Architecture of the AIS for surveillance of the plague.

- age and sex structure of principal host population according to seasons, years, various phases of the population dynamics as well as in the nuclei and peripheral parts of colonies;
- reproduction of the principal host population;
- mortality rate of the principal host population;
- phenotypical structure of the principal host population in different parts of its area and its dynamics in different years and seasons against the background of the changes in the epizootic state in natural foci;
- structure of the principal host populations according to blood groups and its dynamics depending on plague epizootics;
- dynamics of the level of infection susceptibility of the principal host at various phases of the epizootic cycle of a focus;
- assessment of the physiological state of the principal host populations according to seasons, years, and at different phases of the population dynamic;
- correlation of mono- and polygamous families in the principal host population (the large gerbil).

The organism level of research involves obtaining the following information:

- determination of age, sex, and generative state of hosts;
- assessment of the physiological condition of animals of different ages and both sexes;
- level of individual susceptibility to infection in the principal host populations;
- mobility and the migration activity of animals of different ages and sexes in the principal host population.

The data characterizing the state of the members of the epizootic triad are classified according to the season (spring, summer, autumn), and sometimes (for example, in case of reproductive indices) according to months and even decades. The annual and seasonal growth indicating the speed and direction in the population changes are calculated from the basic seasonal indices of the members of the epizootic triad.

Weather, geographical, and space characteristics constitute the indices of *external factors*. The weather characteristics are the temperature of the air, soil, and precipitation; various hydrothermal coefficients (the relation of the amount of precipitation and the sum of temperatures over a certain period of time); the number of days below freezing without thaw; the sum of air temperatures over the period when it was above 10°C, 5°C; the recurrence of wind and the wind velocity according to compass points; the recurrence of the types of atmospheric circulation, the value of flood; etc.

130 Applications

The geophysical indices include the indices of geomagnetic activity, planetary and local, while the space characteristics include Wolf's numbers and the number of patches located along the Sun's central meridian.

6.3.2 AIS for Surveillance of the Plague

Complex interactions as are found in the plague system need new methods and approaches in order to improve the surveillance of epizootic and epidemic situations in the natural plague foci. Development of AIS has the potential to achieve this. The AIS is intended to improve risk analysis and underlying understanding of the space-time dynamics of the plague in Central Asia. Moreover, the AIS could be considered as a part of a general surveillance system for other infectious diseases that are reemerging through tourism and other international activities. Also the AIS could be considered as a model system for processing surveillance data on dangerous infections in all parts of the world.

Consider the architecture of the proposed AIS in Figure 6.6.

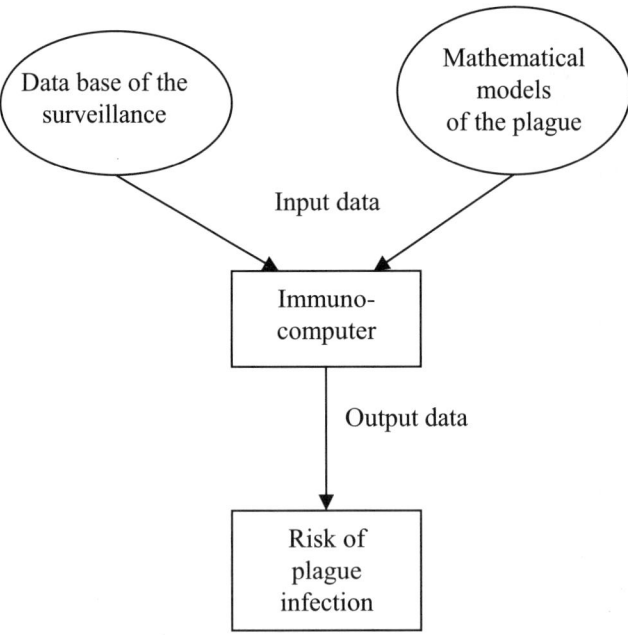

Figure 6.6 Architecture of the AIS for surveillance of the plague.

Input data for the AIS are generated by two main blocks:

1. Database of surveillance of the plague in Central Asia, including computerized existing historic data currently available only in hard copy;
2. A set of mathematical models (stochastic, interval, and discrete) of the space-time dynamics of the plague.

Based on the input data, the main functions of the AIS consist in evaluating the current danger of plague infection, as well as predicting risk of infection in the future. To perform such functions, the core of the AIS represents a pattern recognition block (the software of this block emulates our IC approach to pattern recognition). As (Tarakanov, Sokolova, Abramov, and Dubyansky, 1999) have shown, this approach can be used successfully for complex evaluation of the plague epizootic. Consider now an application of the approach to predict the risk of plague infection (Tarakanov, Sokolova, Abramov, and Aikimbayev, 2000).

Figure 6.7 and Table 6.10 give input data for this task.

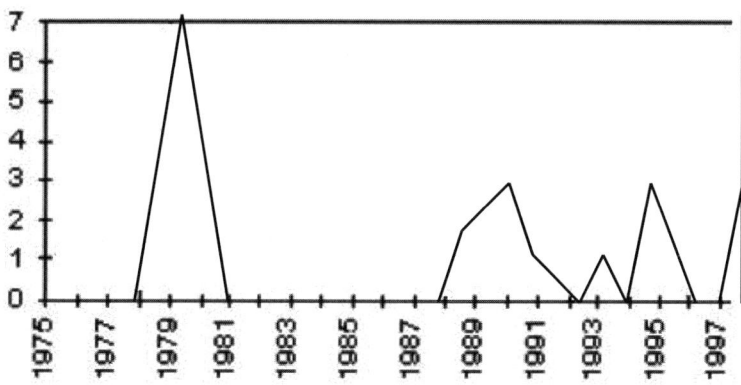

Figure 6.7 Space-time dynamics of the plague on the Akdala plain.

Figure 6.7 shows numbers of infected sectors (vertical axis) over several years (horizontal axis) for the Akdala region. Each sector represents a square of 10×10 kilometers. If the plague's host (rodent), which has been caught in the sector, contains a plague microbe, then the entire sector is considered as infected. It is worth noting that the number of infected sectors obtained by such a method now represents the most appropriate indicator of the epizootic process (Tarakanov, Sokolova, Abramov, and Dubyansky, 1999).

A fragment of the database of surveillance from the Akdala plague focus during 11 years is represented in Table 6.10.

132 Applications

Year→	1976	1977	1978	1979	...	1985	1986	1987	1988
p1	3.3	1.3	5.2	8.8	...	0.4	8.2	4.0	23.5
p2	0.7	4.1	29.6	0.6	...	0.1	4.7	4.4	46.1
p3	3.3	7.1	54.1	425	...	1.9	53.4	170	480
...	...	...	...	...	...	...	...	...	...
p45	5.0	4.0	2.7	2.3	...	5.0	1.7	2.3	1.0

Table 6.10 A fragment of the database of surveillance of the plague.

The fragment includes 45 parameters p1–p45, which are as follows:

- Number of rodents per square: p1 (in autumn), p2 (in spring).
- Number of infected rodents per square: p3 (in autumn), p4 (in spring).
- Average temperature: p5 (March) – p10 (September).
- Average humidity (%): p11 (March) – p16 (September).
- Average temperature of the soil at a depth of 0.2 meters:
  p17 (March) – p22 (September).
- Average temperature of the soil at a depth of 1.6 meters:
  p23 (March) – p28 (September).
- Average wind speed: p29 (March) – p34 (September).
- Total atmospheric precipitates: p35 (March) – p40 (September).
- Average height of the snow blanket:
  p41 (January) – p43 (March), p44 (November) – p45 (December).

Now consider the task of predicting the risk of infection (indicator in Figure 6.7) from the data in Table 6.10.

6.3.3 Supervised Learning

Input data for this task are given by the columns of Table 6.10 and by the second column of Table 6.11. Experts, according to Figure 6.7, had assigned the classes for this column. Thus, learning patterns are given by the years 1976–1981, and the task is to assign the corresponding classes to the years 1984–1988. Classes that have been assigned by the AIS to the years 1976–1981 are considered as a test of the learning process.

According to the IC approach the task is solved as follows:

Folding Vectors to Matrices

Fold every column R_{year} of Table 6.10 (vector of the dimension 45×1) to a matrix A_{year} of dimension 9×5. It has been shown, in Section 3.1, that such folding increases the precision (specificity) of recognition.

Year	Class by experts	$-w_1$	$-w_2$	$-w_3$	Class by AIS	Risk of infection
1976	1	<u>144</u>	14	143	1	mid
1977	1	<u>133</u>	18	130	1	mid
1978	1	<u>159</u>	72	158	1	mid
1979	2	180	<u>432</u>	174	2	high
1980	3	146	23	<u>149</u>	3	low
1981	3	150	15	<u>150</u>	3	low
1984		<u>150</u>	11	147	1	mid
1985		146	13	<u>148</u>	3	low
1986		<u>152</u>	66	150	1	mid
1987		174	<u>182</u>	174	2	high
1988		197	<u>499</u>	180	2	High

Table 6.11 Results of recognition.

Learning

Form matrices A_1, A_2, A_3 for classes 1,2,3, respectively:

$$A_1 = \frac{A_{1977} + A_{1978}}{2}, \quad A_2 = A_{1979}, \quad A_1 = \frac{A_{1977} + A_{1978}}{2}.$$

Compute their singular vectors by the SVD:

$$\{X_1, Y_1\} \text{ for } A_1,$$
$$\{X_2, Y_2\} \text{ for } A_2,$$
$$\{X_3, Y_3\} \text{ for } A_3.$$

Recognition

Compute three values of the binding energy for every input pattern A_{year}:

$$w_1 = -X_1^T A_{\text{year}} Y_1,$$
$$w_2 = -X_2^T A_{\text{year}} Y_2,$$
$$w_3 = -X_3^T A_{\text{year}} Y_3.$$

Determine the class to be found by the minimal value of the binding energy:

134 Applications

$$k : w_k = \min_k \{w_1, w_2, w_3\}.$$

The results of the recognition are shown in Table 6.11, where the minimal values of the energy are underlined. According to Figure 6.7, the classes that have been recognized could also be treated in the last column of Table 6.11 as a risk level of plague infection.

6.3.4 Unsupervised Learning

Consider a matrix A of dimension 11×45 formed by all the input data from Table 6.10. Compute the SVD of the matrix, where s_1, s_2 are the first two singular values, and Y_1, Y_2 are the right singular vectors corresponding to these values:

$$A = s_1 \begin{bmatrix} x1_1 \\ \ldots \\ x1_{11} \end{bmatrix} Y_1^T + s_2 \begin{bmatrix} x2_1 \\ \ldots \\ x2_{11} \end{bmatrix} Y_2^T + \cdots.$$

According to Section 3.1, a rigorous correspondence can be established between vectors and FPs. So, consider two FPs as antibodies: {FP-1, FP-2}, which correspond to vectors Y_1, Y_2. Also consider eleven FPs: {FP$_1$, ... , FP$_{11}$}, which correspond to the strings of matrix A (columns of Table 6.10). Then every string A_i, which represents the number of a specific year $i = 1,...,11$, can be mapped to the 2 values $\{x1_i, x2_i\}$ of the binding energy between FP$_i$ and two antibodies:

$$x1_i = w(\text{FP-1}, \text{FP}_i), \quad x2_i = w(\text{FP-2}, \text{FP}_i).$$

The results are given in Table 6.12 and are represented geometrically in Figure 6.8, where the framed group of the close years in the right-bottom corner is enlarged for a clearer view.

Every year is represented in Figure 6.8 by a point in a 2D shape space of the binding energies $\{x1, x2\}$.

Figure 6.8 clearly detects three groups (classes) of the years with relatively similar complex conditions according to the number of hosts and condition of the weather:

- class A = {1979, 1988},
- class B = {1987},
- class C = {all other years}.

Year	x1	x2	Class	Risk of infection
1976	0.106	0.335	C2	
1977	0.101	0.297	C3	
1978	0.184	0.263	C1	
1979	0.560	−0.211	A	high
1980	0.125	0.313	C2	
1981	0.108	0.339	C3	
1984	0.107	0.356	C4	
1985	0.109	0.353	C4	
1986	0.170	0.272	C1	
1987	0.298	0.153	B	mid
1988	0.680	−0.352	A	high

Table 6.12 Classification of the years by a shape space of the AIS.

Figure 6.8 clearly detects three groups (classes) of years with relatively similar complex conditions according to the number of the hosts and condition of the weather:

- class A = {1979, 1988},
- class B = {1987},
- class C = {all other years}.

Finer analysis within class C allows for selection of the following subclasses of the years with very close complex conditions:

- C1 = {1978, 1986},
- C2 = {1976, 1980},
- C3 = {1977, 1981},
- C4 = {1984,1985).

Comparison of this classification with the real indicator of the infection in Figure 6.7 allows us to connect the position of the point on the shape space with the risk of infection. Namely, risk increases from the right-bottom corner in Figure 6.8 (class C) to the left-top corner (class A). Moreover, class A = {1979, 1988} sharply contrasts with all other classes in Figure 6.8. Therefore, if the complex condition of hosts and weather for any year is similar to those of class A, then the condition gives a high risk of plague infection.

136 Applications

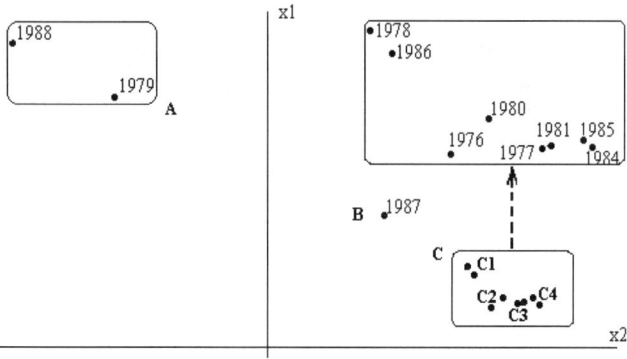

Figure 6.8 Number of hosts and condition of the weather on a shape space.

6.3.5 Comparison with Traditional Statistics

Consider the correlation coefficients in Table 6.13, which are computed by the time series of number of hosts and condition of the weather (Table 6.10). High correlations between the years with coefficient greater than 0.9 are underlined in Table 6.13.

Year	1977	1978	1979	1980	1981	1984	1985	1986	1987	1988
1976	<u>96</u>	76	17	80	<u>91</u>	<u>96</u>	<u>92</u>	87	51	07
1977		77	19	79	<u>94</u>	<u>97</u>	88	85	51	10
1978			50	84	73	71	72	88	74	55
1979				18	15	14	15	53	<u>91</u>	90
1980					87	80	85	83	53	21
1981						<u>94</u>	87	82	50	05
1984							90	84	48	05
1985								87	51	04
1986									79	49
1987										78

Table 6.13 Correlation coefficients (multiplied by 100) between the time series of Table 6.10.

Even a brief survey of Table 6.13 shows that the most dangerous correspondence (between the years 1979 and 1988 with a high risk of infection) is not as evident as with the IC approach in Figure 6.8. At the same time, all the years with high correlation are relatively close on the shape space of the IC.

Therefore, it can be concluded that the IC is able to focus sharp attention on the most dangerous situations, which is beyond the possibilities of traditional statistics.

6.3.6 Conclusions

The results presented in Section 6.3 show that IC is a powerful, robust; and flexible approach with which to process the data of surveillance of natural plague foci. It could be taken as a model approach for surveillance of similar infectious diseases in all parts of the world.

Another promising direction of such surveillance could be combining of GIS technology with remote sensing data of satellites (Kitron, 1998). Here IC provides a tool for analyzing and integrating the heterogeneous data of epidemiology, ecology, and the space-time dynamics of infection.

6.4 Intelligent Security Systems

The protection of people and property is one of the main problems always faced by society. Full responsibility for this protection and for the prevention of criminal activity has been taken by governments (through its departments of justice and law enforcement). This responsibility involves the protection of individuals, industrial enterprises, banks, shopping centers (both small and large), cities, etc.

Security systems play an important role in the prevention of the above-mentioned property crimes. These systems have all the features of complex systems:

- large numbers of interconnected elements;
- information uncertainty on potential criminal activity and its perpetrators;
- involvement of "human factors" that manifest themselves in decision-making;
- natural factors, e.g., a wide variety of climatic conditions;
- natural and industrial disturbances.

It is well known (C and K Systems, 1997; Lavrus, 1996) that approximately 95% of generated alarms are false alarms, caused by malfunctions of equipment and environmental noise. In most cases, the particular sources of false alarms are difficult to detect.

An IC approach for intelligent analysis of information generated by multisensor security systems has been developed, and an intelligent security

138 Applications

system (ISS) for complex objects has been proposed (Sokolova et al., 2000), greatly enhancing the dependability of such systems.

This ISS ensures the realization of the following functions:

- assessment of the status of complex objects and the analysis of abnormal surveillance information;
- implementation of procedures of supervised and unsupervised learning;
- implementation of pattern recognition procedures for abnormal surveillance information in the shape space of IC.

At this time, passive infrared (PIR) detectors are widely utilized, providing one of the basic means for protection of designated areas, roadways, etc. The PIR technology is known as "passive" due to the absence of any radiation, which makes these detectors absolutely harmless to the people in the area of surveillance. This technology can also be used for protection of fragile materials, e.g., in museums. The principle of operation of a PIR-detector is based on measurement of a difference in temperatures between the intruder's body and environmental surfaces (walls, floors, and furniture) within the visibility range of the receiving optics of the device.

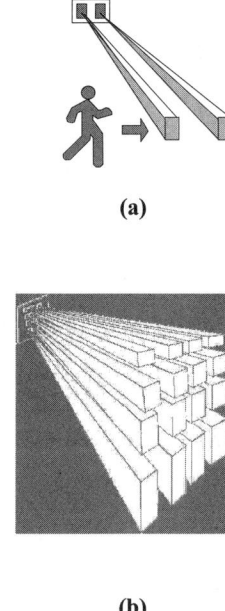

(a)

(b)

Figure 6.9 (a) Separation of information channels in PIR-detectors; (b) an example of a possible zone of detection of the PIR-detector with a 4×4 matrix.

6.4 Intelligent Security Systems

The major shortcoming of PIR-detectors is their inability to localize the intrusion area and determine the size of the intruding body. Localization of the place of intrusion is necessary for adequate reaction by the various enforcement agencies. Correct estimation of the body size of the intruder can greatly reduce the number of false alarms, since the source of the alarm could easily be a small animal (cat, dog, or rabbit).

To eliminate this shortcoming, a dedicated information channel has been created for each sensor of the PIR-detector, creating 4×4, 8×8, 16×16 matrices of sensitive elements. This facilitates the localization of intrusions and reduces the malfunctioning of the entire system. The provisional zone of detection of the PIR-detector with a 4×4 matrix of sensitive elements is given in Figure 6.9. In this device it has four vertical layers with four beams in each layer. Consider the ways of detecting an intruder in situations (a) and (b) in the diagram. Initially, all sensitive elements are in a normal state, defined as a_0, which could be changed into a_1 by contact with an intruder registering a change of condition of the protected object.

If the intruder's body size is small and comparable with the size of a beam (for example a cat or a dog), there will be a serial operation of separate sensitive elements whose condition is registered in a buffer device and transferred to a computer for processing. The change of conditions is shown in Table 6.14.

Phase 1	Phase 2	Phase 3	Phase 4
$a_0\ a_0\ a_0\ a_0$	$a_0\ a_0\ a_0\ a_0$	$a_0\ a_0\ a_0\ a_0$	$a_0\ a_0\ a_0\ a_0$
$a_0\ a_0\ a_0\ a_0$	$a_0\ a_0\ a_0\ a_0$	$a_0\ a_0\ a_0\ a_0$	$a_0\ a_0\ a_0\ a_0$
$a_0\ a_0\ a_0\ a_0$	$a_0\ a_0\ a_0\ a_0$	$a_0\ a_0\ a_0\ a_0$	$a_0\ a_0\ a_0\ a_0$
$a_1\ a_0\ a_0\ a_0$	$a_0\ a_1\ a_0\ a_0$	$a_0\ a_0\ a_1\ a_0$	$a_0\ a_0\ a_0\ a_1$

Table 6.14 The order of operations of sensitive elements during movement of a small intruder.

Phase 1	Phase 2	Phase 3	Phase 4
$a_0\ a_0\ a_0\ a_0$	$a_0\ a_0\ a_0\ a_0$	$a_0\ a_0\ a_0\ a_0$	$a_0\ a_0\ a_0\ a_0$
$a_0\ a_0\ a_0\ a_0$	$a_0\ a_0\ a_0\ a_0$	$a_0\ a_0\ a_0\ a_0$	$a_0\ a_0\ a_0\ a_0$
$a_1\ a_0\ a_0\ a_0$	$a_0\ a_1\ a_0\ a_0$	$a_0\ a_0\ a_1\ a_0$	$a_0\ a_0\ a_0\ a_1$
$a_1\ a_0\ a_0\ a_0$	$a_0\ a_1\ a_0\ a_0$	$a_0\ a_0\ a_1\ a_0$	$a_0\ a_0\ a_0\ a_1$

Table 6.15 The order of operations of sensitive elements during movement by a large intruder.

Depending on the means used to detect an intrusion, factors a_0 and a_1 may be assigned the appropriate Boolean values, 0 or 1, consistent with threshold

140 Applications

algorithms, or integer values within any fixed interval (for example, in a range from 0 up to 255 for application of 8-bit analog–digital converters).

In Table 6.15 the operation of sensitive elements is consistent with the movement of a large intruder.

Even preliminary analysis shows the obvious advantages of providing each sensitive element an individual information channel. In schemes that utilize this option, the following additional opportunities could be observed:

- estimation of the size of the intrusion zone (the intruder);
- definition of a number of activated sensitive elements, and therefore the place of intrusion;
- direction of movement of the intruder.

Application of PIR-detectors with individual information channels allows for detection of the paths of intrusion into a protected zone. For example, in the case of intrusion through a ceiling, the sensitive elements will be activated in a specific sequence, first the fourth, upper, layer, then the third layer, etc. Using this information, law enforcers may evaluate the tactics of intrusions.

The quality of detectors with individual information channels may be improved in the following two ways:

1. by increasing the number of elements in a matrix of the PIR detector;
2. by the application of spatial, mechanical, or optical scanning.

Assume that the mode and localization of an intrusion have been determined with the help of a PIR-detector with a 4×4 matrix and individual information channels. Assume that the timing of activation of particular sensor devices by the PIR-detector is known to be

$$\widetilde{X}_1 = [1, 0.1, 0.4, 1, 1, 1, 0.2, 1, 0.1, 1, 1, 0.7, 0.3, 0.8, 1, 0.2]^T,$$
$$\widetilde{X}_2 = [0.1, 0.6, 0.2, 1, 1, 1, 0.4, 1, 0.1, 1, 1, 0.7, 1, 0.8, 1, 1]^T,$$
$$\widetilde{X}_k = [1, 0.1, 0.8, 1, 1, 0.5, 1, 1, 0.1, 1, 1, 0.4, 0.2, 0.4, 1, 0.5]^T.$$

Let us treat an arbitrary vector of indicators $\widetilde{X}_k$, $k=1,...,n$, as a pattern of the indicator space $\{X\}$, and a set of patterns to be represented as a set of vectors consisting of k subsets $z_1=\{X\}_1,..., z_k=\{X\}_k$.

Assume that this information is presented to an expert, who classifies it on the basis of personal experience.

On the basis of the obtained information and consistent with the expert's judgment, a learning sample of surveillance data could be formed (Table 6.16).

Number of the situation	Values of indicators					Class by the expert
	z_1	z_2	z_3	...	z_{16}	
1	1	0, 3	1		1	1
2	0, 1	0, 6	1		1	2
3	0, 2	0, 8	0, 9		0, 7	3
4	0, 5	1	0, 7		0, 1	4
...					...	
:						
n	1	1	0, 5		0, 1	1

Table 6.16 Learning sample by results of analysis of abnormal situations

The task of learning is reduced to a partition of the indicator space into classes (classification), and the task of pattern recognition is reduced to the definition of the class $z_j = \{\widetilde{X}\}_j, j = 1,...,k,$ with the help of the vector norm:

$$z_k(\widetilde{X}): \min_k \|\widetilde{X} - \{X\}_k\|.$$

The tasks of learning, pattern recognition, and automatic classification of abnormal surveillance data could be performed on the basis of the IC approach by developing an ISS for complex objects. The structure of this system includes a knowledge base, an information database, procedures of supervised learning (with an expert), unsupervised learning (without an expert), pattern recognition, and automatic classification of the abnormal surveillance data.

6.4.1 Supervised Learning

The procedure of supervised learning consists of the following steps:

Folding a Vector to a Matrix

The initial data set is represented by a matrix, which determines the binding energy between FPs:

$$\widetilde{X} = [x_1,...,x_n]^T \to A(\widetilde{X}).$$

For example, the vector of indicators $\widetilde{X}_1$ of dimension 16×1 of the form

$$\widetilde{X}_1 = [1, 0.1, 0.4, 1, 1, 1, 0.2, 1, 0.1, 1, 1, 0.7, 0.3, 0.8, 1, 0.2]^T$$

is folded into a matrix of dimension 4×4:

$$A_1 = \begin{bmatrix} 1 & 0.1 & 0.4 & 1 \\ 1 & 1 & 0{,}2 & 1 \\ 0.1 & 1 & 1 & 0.7 \\ 0.3 & 0.8 & 1 & 0.2 \end{bmatrix}.$$

SVD of the Matrix

For a matrix A_1 the groups of right and left singular vectors and singular values, respectively, are

$$L_1 = \begin{bmatrix} 0.449 \\ 0.596 \\ 0.521 \\ 0.414 \end{bmatrix}, \quad R_1 = \begin{bmatrix} 0.447 \\ 0.546 \\ 0.451 \\ 0.546 \end{bmatrix},$$

$$L_2 = \begin{bmatrix} -0.532 \\ -0.375 \\ -0.567 \\ 0.505 \end{bmatrix}, \quad R_2 = \begin{bmatrix} -0.578 \\ 0.373 \\ 0.622 \\ -0.374 \end{bmatrix},$$

$$L_3 = \begin{bmatrix} 0.008 \\ -0.020 \\ 0.999 \\ -0.195 \end{bmatrix}, \quad R_3 = \begin{bmatrix} -0.648 \\ 0.329 \\ 0.643 \\ -0.241 \end{bmatrix},$$

$$s_1 = 2.7337,$$
$$s_2 = 1.2824,$$
$$s_3 = 1.4688.$$

The singular value decomposition for matrices A_k, $k=1,...,n,$ is also carried out. Appropriate groups of right and left singular vectors and singular values are also determined.

6.4 Intelligent Security Systems

Number of an alarming situation	Values of indicators				Binding energy				Membership in the class
	z_1	z_2	...	z_{16}	w_1	w_2	w_3	w_4	
10	0.2	0.8		1	6.2	5.2	4.2	3.2	4
11	1	0.1		0.7	3.1	1.9	2.0	1.6	4
12	1	1		0.5	1.0	0.9	1.6	2.4	2
13	0.6	0.1		1	3.8	1.8	2.6	3.7	2
N	0.2	1		1	10.5	10.8	8.5	9.3	3

Table 6.17 Outcome of the analysis.

Recognition

For each estimated vector $\widetilde{X}$, the class $\{L_i, R_i\}_k$ with a minimum binding energy is determined according to the minimal distance (3.17). The results of pattern recognition obtained on the basis of the procedure of supervised learning are presented in Table 6.17.

6.4.2 Unsupervised Learning

The procedure of unsupervised learning (without an expert) is based on automatic classification. A group of classes is computed as functions of norms of singular vectors of a total matrix, which includes all vector data:

$$A = A(\{X\}) \Rightarrow \sum_{i \geq I}^{\leq p} \|(R_i)_k\| \Rightarrow z_k.$$

For example, we shall consider a matrix of dimension 11×16, formed by the input data of Table 6.14 (including the first column). Let us compute the SVD of this matrix:

$$A = s_1 \begin{bmatrix} x1_1 \\ \ldots \\ x1_{11} \end{bmatrix} Y_1^T + s_2 \begin{bmatrix} x2_1 \\ \ldots \\ x2_{11} \end{bmatrix} Y_2^T + \cdots,$$

where s_1, s_2 are the first two singular values, Y_1, Y_2 are right singular vectors.

144 Applications

Each row A_i, $i = 1, ..., 11$, of the matrix A represents the number of abnormal surveillance data samples (situations) and can be described by two values of binding energies $w_1 = x1$, $w_2 = x2$ between Y_1^T, Y_2^T and A_i respectively:

$$x1_i = w(Y_1, A_i), \quad x2_i = w(Y_2, A_i).$$

The outcomes are reduced in Table 6.18 and geometrically represented in Figure 6.10, where the chosen group of abnormal situations in the lower right corner is enlarged for clarity.

Situation #	Binding energy w_1	Binding energy w_2	The membership class	Degree of danger
1	1, 2	−1, 3	1	High
2	1, 5	−1, 5	1	High
3	1, 1	1, 2	2	Average
4	0, 7	0, 1	3	Low
...	1, 21	1, 3	2	Average
100	1, 5	−1, 19	1	High
...				

Table 6.18 Classification of abnormal situations.

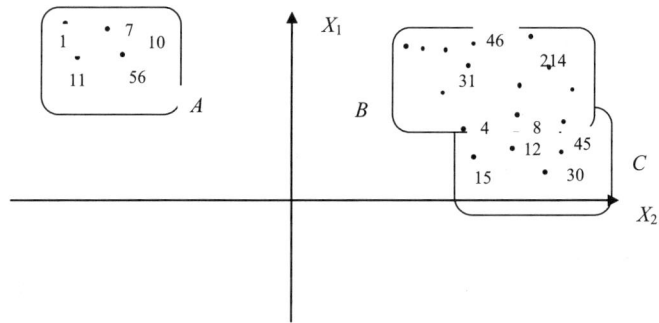

Figure 6.10 Classification of abnormal situations in 2D shape space.

The intelligent security system has allowed for learning on the basis of the available abnormal surveillance data, in order to perform pattern recognition and to achieve classification of current abnormal data in the shape space of IC.

7
Immunocomputing Systems: Architecture and Implementation

7.1 Toward an Immunocomputer

In this section we discuss the concept of an *immunocomputer* (ICr) as a computer "hard wired" to implement immunocomputing (IC), i.e., utilizing a computational approach that reflects the principles of information processing executed by proteins and immune networks. The main purpose of the ICr is to perform high-throughput computing according to the IC approach in a parallel and distributed manner, as in the natural immune system. In previous chapters we have developed a proper mathematical basis for the ICr by introducing the notions FP and FIN. The need for these notions is prompted by the very specific objects and interactions of immune networks, which are remarkably different from those of genetic algorithms, cellular automata, ANN, and intelligent agents.

We believe that such a mathematical basis could raise AIS to the same level as the widespread ANN, and even allow us to speak about hardware implementation of FIN in a special *immunochip*. Such a chip could be treated as the core of a future ICr.

We also present some potential applications of the immunochip to information security, including intrusion detection, steganography, and cryptography.

7.1.1 Immunochip Architecture

In addition to proteins, cells—which produce and secrete proteins, and also employ proteins as their receptors—can be considered the second basic component of information processing by immune networks. Two main sorts of immune cells can be distinguished: B-cells and T-cells.

Let us distinguish two kinds of proteins: "free proteins," independent of cells, and proteins anchored in cell membranes as receptors. Examples of free proteins are *peptides* (small proteins), antigens, antibodies produced by B-cells, and numerical peptides (*lymphokines*) produced by T-cells. Examples of receptors are the proteins known as MHC I and MHC II (*major histocompatibility complex* class I and II). These proteins are used by the immune system as universal markers of the body's own cells (self-cells) to distinguish them from nonself antigens.

The architecture of any computer also includes at least two basic components: memory and processor. They can be gathered in separate modules, like RAM (random-access memory) and the CPU (central processing unit) in the traditional PC, or distributed among other structural elements, like the neurons of neurocomputers, or cells of cellular automata. Nevertheless, memory and processing units are the intrinsic components of any computer.

Thus, consider the architecture of an immunochip as shown in Figure 7.1.

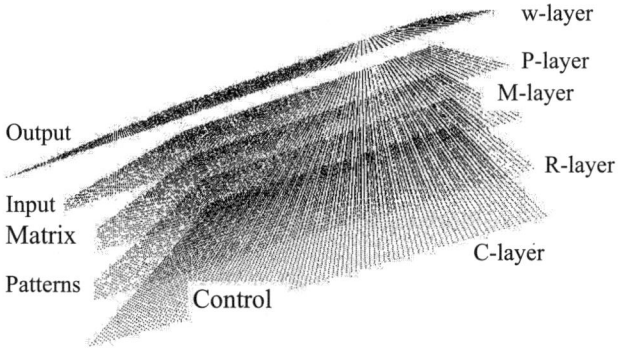

Figure 7.1 Architecture of an immunochip.

Each memory cell is depicted in Figure 7.1 as a point. Cells are gathered in arrays, depicted in Figure 7.1 as layers (top-down):

1. Output array (w-layer), whose cells contain values of a bilinear form that correspond to the binding energy between the cells of the P-layer and R-layer;
2. Input array (P-layer), or sample array, in which every cell contains one vector to be recognized;
3. Intermediate array (M-layer), which contains a matrix to determine the bilinear form (binding energy);
4. Array of probes (R-layer), whose cells store unit vectors as patterns;
5. Control array (C-layer), whose cells contain a set of parameters for the changing of stored patterns.

These hierarchical layers are intended for storing real values in a floating-point format. Accordingly, each cell of the w-layer stores one value, and cells of the P-layer, R-layer, and C-layer store vectors of similar values. Cells of the M-layer store matrices of similar values. The processing units of the immunochip (not shown in Figure 7.1) compute a bilinear form using cells of the P-layer, R-layer, and M-layer and put the result in the cells of the w-layer. Processing units also change cells of the R-layer and C-layer using a control set stored in the cells of the C-layer.

Assume that every memory unit has only well-defined neighboring units. Namely, every memory unit has:

1. Four vertical neighbors arranged above and/or below the unit in any adjacent array;
2. Four horizontal neighbors perpendicularly crossed in the same array, as shown in Figure 7.2.

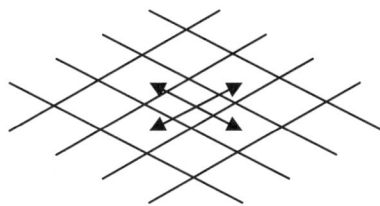

Figure 7.2 Four horizontal neighbors of a memory cell.

Consider the content of any memory cell as its *state*. Then the general function of the immunochip is to determine the states of the output array by the states of the input array, in accordance with stored patterns that can change dynamically. For this purpose, the processing units of the immunochip only determine interactions between states of the neighboring units.

It is worth noting that the vertical interactions in such an immunochip correspond to already existing *biochips*, also called *microarrays* (Ekins and Chu, 1999; MacBeath and Schreiber, 2000). In fact, the microarray of a biochip is an orderly arrangement of probes, such as short strands of DNA (*deoxyribonucleic acid*) or antibodies (corresponding to the R-layer of the immunochip). These probes are immobilized in a solid surface, such as nylon, glass, or silicon substrates (C-layer), and exposed to a set of testing samples (P-layer). The results of binding between samples and probes are determined by fluorescence or electric signals (w-layer).

On the other hand, if any memory array of the immunochip is able to store only a few discrete states, and all units of the array change states simultaneously in discrete time, then the well-known cellular automata machines (Toffoli and

Margolus, 1987) or excitable media (Adamatzky, 1997) could realize horizontal interactions within the array.

However, such special kinds of immunochip are obviously insufficient to simulate features of immune networks. Hence, consider that any memory unit of the immunochip is able to store a set of real numbers, and processing units are able to compute this set using the set of any horizontal or vertical neighbor.

To illustrate the immunochip architecture, consider an implementation of a mathematical model of AIS (Tarakanov and Dasgupta, 2000). Let the R-layer contain unit vectors R as a model of the shapes of antibodies. Let an input flow of unit vectors P be given in the P-layer as a model of a biomolecular environment. Let each cell of the M-layer contain the same matrix M that determines the binding energy w between corresponding vertical neighbors P and R and puts the value of w in the cell that is the vertical neighbor of the w-layer. Let an antibody recognize an antigen if the value of w is lower than the threshold given in the vertical neighbor cell of the C-layer. If such recognition occurs, consider that all horizontal neighbors of the antibody R are replaced by the slightly modified vector R, according to a mutation rate given in the cell of the C-layer. This immunochip model is thus able to simulate primary and secondary immune responses in the form of total changes in antibody population of the R-layer. Consider the dimensions of the layers as $10^4 \times 10^4$ cells and the vector R as a 12-dimensional vector with 10 discrete values of each component. Then the immunochip implementation is able to contain 10^8 antibodies of 10^{12} potential shapes. According to (Segel and Cohen, 2001), such numbers are close to those of the natural immune system.

7.1.2 Implementation of Mathematical Operations

Let the states of the memory cells be w_{ij}, P_{ij}, M, R_{ij}, C_{ij}, where i and j are row and column numbers (addresses) of cells within the array, w_{ij} is a real value of binding energy, P_{ij}, R_{ij}, and C_{ij} are vectors with real components of dimensions n_P, n_R, and n_C, respectively, and M is a matrix with real components with dimensions $n_P \times n_R$. Consider vectors P_{ij}, R_{ij}, and C_{ij} as column vectors that code the states of proteins, receptors, and cells, respectively.

Let the binding energy be defined by a bilinear form

$$w_{ij} = -P_{ij}^T M R_{ij}. \tag{7.1}$$

Now let us consider several special cases implementing important mathematical operations.

Singular Value Decomposition

Assume that the matrix M is given. Consider a set of unit vectors P_{ij} and R_{ij} with dimensions n_P and n_R, respectively:

$$P_{ij}^T P_{ij} = R_{ij}^T R_{ij} = 1, \ \forall i,j.$$

Compute w_{ij} for all pairs of such vectors as in (7.1). Select the minimal $w_{ij}=w^*$ and the corresponding pair of vectors P^*, R^*:

$$w^* = \min_{i,j}\{w_{ij}\}, \ w^* = -[P^*]^T M R^*. \qquad (7.2)$$

Let the value $w^*=w(P^*, R^*)$ satisfy the following condition:

$$w^* \leq w(P,R), \ \forall P,R : P^T P = R^T R = 1.$$

Then, according to (3.5), $s_1=-w^*$ is the maximal singular value of the matrix M, while $X_1=P^*$ and $Y_1=R^*$ are the left and the right singular vectors of this matrix.

Compute the matrices M_k by the following recurrent rule:

$$M_k = M_{k-1} - s_{k-1} X_{k-1} Y_{k-1}^T,$$
$$k = 2,...,r,$$
$$M_1 = M,$$

where r is the rank of matrix M.

Similarly, determine the maximal value $s_k=-w^*$ and the corresponding vectors $X_k=P^*$ and $Y_k=R^*$ for the matrix M_k.

Finally, we obtain the SVD of the initial matrix M:

$$M = s_1 X_1 Y_1^T + \cdots + s_r X_r Y_r^T.$$

Note that the immunochip allows for minimization of the value w_{ij} in at least two different ways, and is thus close to evolutionary computation.

First, the immunochip can use a process of random "mutations" of vector coordinates so that P_{ij} and R_{ij} still remain unit vectors. For example, if the immunochip has received a value of w_{ij} that satisfies (7.2), and this value has not been reduced after a large number of mutations, then this value could be considered a minimum.

Second, the immunochip is able to use a more rigid procedure. For example, let w_{ij}, P_{ij}, R_{ij}, and C_{ij} be computed by the following recurrent scheme (we omit the subscripts for convenience):

$$[R^{(k)}]^T = [P^{(k-1)}]^T M,$$
$$C^{(k)} = MR^{(k)},$$
$$P^{(k)} = C^{(k)},$$
$$w^{(k)} = -[P^{(k)}]^T MR^{(k)},$$
$$k = 2,...,$$

while $\left|w^{(k)} - w^{(k-1)}\right| > \varepsilon$.

According to (Horn and Johnson, 1986), such a scheme converges to the maximal singular value and singular vectors in the general case of the matrix M.

Formal Immune Networks

Consider two-dimensional unit vectors. Then any vector can be represented as depending on one angle, for example,

$$S(\varphi) = [\cos\varphi, \sin\varphi]^T.$$

Let this angle have only one of n discrete values:

$$\varphi_1 = \frac{2\pi}{n},$$
$$\varphi_k = k\varphi_1,$$
$$k = 0, 1, ..., n-1.$$

This leads to exactly n types of the vectors. Designate them as $S^{(k)}$, $k=0, ..., n-1$.

Define the matrix M as the identity matrix

$$M = \begin{bmatrix} 1 & 0 \\ 0 & 1 \end{bmatrix}.$$

Then the binding energy between vectors $S^{(k_1)}$ and $S^{(k_2)}$ is

$$w = -\cos(\varphi_{k_1} - \varphi_{k_2}). \tag{7.3}$$

Let an integer n_h define the threshold of binding w_h as follows:

$$w_h = -\cos(n_h \varphi_1).$$

Hence, the binding condition can be reduced to the following inequality:

$$\min\{|k_1 - k_2|,\ n - |k_1 - k_2|\} \le n_h.$$

Let the memory of the immunochip be one-dimensional. Then the states can be marked by an index j and represented in the form of a matrix:

$$\begin{bmatrix} w_1 & \ldots & w_j & \ldots \\ P_1 & \ldots & P_j & \ldots \\ R_1 & \ldots & R_j & \ldots \\ C_1 & \ldots & C_j & \ldots \end{bmatrix}.$$

Designate an empty memory unit (*gap*) by the symbol $\emptyset$. Let the initial sequence (*population*) $\{R\}$ of length m without gaps be given:

$$\{R : (R_j \ne \emptyset\ \forall j \le m),\ (R_j = \emptyset\ \forall j > m)\}. \tag{7.4}$$

Let the population $\{P\}$ be arbitrarily determined, and the initial population $\{C\}$ be empty. Consider processing population $\{R\}$ by the following algorithm:

Algorithm 7.1

1. Compute w_j between P_j and R_j by (7.3);
2. Change R_j and C_j according to w_j and w_h (see rules of step 2 below);
3. Merge sequences $\{R\}$ and $\{C\}$;
4. Repeat steps 1–3 until $\{R\}$ becomes empty or overflows a memory limit.

Step 2 is performed simultaneously for all j by the following rules:

2.1. If $w_j > w_h$ or $P_j = \emptyset$, then $R_j = \emptyset$.
2.2. If $w_j = -1$, then $C_j = R_j$.
2.3. If $-1 < w_j \le w_h$, then

$$C_j = [\text{Rot}(\varphi_1)] R_j,\ R_j = [\text{Rot}(-\varphi_1)] R_j,$$

$$\text{Rot}(\varphi_1) = \begin{bmatrix} \cos\varphi_1 & -\sin\varphi_1 \\ \sin\varphi_1 & \cos\varphi_1 \end{bmatrix}.$$

Simply put, if P_j does not bind with R_j, then R_j either *dies* or *reproduces*. If the strength of the binding is the highest possible, then R_j creates a copy. Otherwise, R_j mutates into the two nearest types.

Step 3 includes the following substeps:

3.1. Attach sequence $\{C\}$ to the end of sequence $\{R\}$;
3.2. If $R_k=\varnothing$ for any k, then shift $R_{j-1}=R_j$ for any $j>k$;
3.3. Perform step 3.2 until (7.4) is satisfied;
3.4. Compute the length m of the resulting sequence $\{R\}$;
3.5. Make sequence $\{C\}$ empty.

According to Section 2.3, Algorithm 7.1 implements the simplest version of FIN, namely the so-called $1DAB(n, n_h)$ network, where sequence $\{P\}$ corresponds to antigens and sequence $\{C\}$ corresponds to B-cells. Several mathematical results can be obtained by the immunochip for such networks (see Section 2.3).

Now consider processing the initial population $\{R\}$ with an empty initial population $\{C\}$ by another algorithm.

Algorithm 7.2

1. Form the sequence $\{P: P_{j-1}=R_j, j=2, ..., m\}$;
2. Compute w_j between P_j and R_j by (7.3);
3. Change the values of R_j and C_j according to w_j and w_h (see rules of step 3 below);
4. Merge $\{R\}\{C\}$ (see step 3 of the Algorithm 7.1);
5. Repeat steps 1–4 until $\{R\}$ becomes empty or overflows the memory limit.

Step 3 is performed simultaneously for all j by the following rules:

3.1. If $w_j>w_h$ or $P_j=\varnothing$, then $R_j=\varnothing$.
3.2. If $w_j\leq w_h$, then

$$C_j=[Rot(\varphi(1))]R_j, \; R_j=[Rot(-\varphi(1))]R_j.$$

Simply put, if P_j does not bind with R_j, then R_j either dies or reproduces with mutations.

According to Section 2.3, Algorithm 7.2 implements another version of FIN, the so-called $1DBB(n, n_h)$ network, where several types of B-cells are generated and stored through interactions among themselves, in spite of the absence of any antigen. Computer simulation shows that variants of cyclic modes in such BB-networks exist, and they have several periods and lengths of populations, including those where the number of B-cells changes from population to population. Apparently, these modes represent the simplest mathematical model of immune memory.

A one-dimensional FIN still yields to pure mathematics. However, if it is two-dimensional, FIN is increasingly fuzzy, and investigation of its properties is practically impossible without computer simulation. Simultaneously, the properties of such FIN seem to be closer to those of the natural immune system. Unsurprisingly, recently used biochips are also two-dimensional. Therefore, immunochips could provide high-throughput simulation of two- and even three-dimensional FINs in a parallel and distributed manner

On the other hand, the mathematical basis of an immunochip relies upon the notions of FP and FIN. According to Sections 2.1 and 2.2, the features of FP allow us to remain closer to the concept of a natural protein, whereas the concept of the artificial neuron has moved away from its biological prototype. In any case, an immunochip provides a more promising model of the natural immune system than a neurocomputer or even cellular automata with discrete states.

Thus, the immunochip could also be a useful device for simulating the natural immune system in connection with such deadly diseases as AIDS. This simulation is essentially based on the hardware implementation of FIN. As shown in Section 2.3, even the simplest variants of FIN possess the inherent properties of immune response and immune memory.

Formal Grammars

Consider any vector as a coded FP, according to Section 3.1. Let $w_h = -1$. Then, according to (7.3), any FP $S^{(k)}$, $k=0,...,n-1$, can bind only with an FP of the same type.

Consider the following initial sequences $\{P\}$, $\{R\}$, and $\{C\}$:

$$\{P : (P_j \neq \varnothing, \; \forall j \leq n), \; (P_j = \varnothing, \; \forall j > n)\},$$
$$\{R : R_0 \neq \varnothing, \; R_j = \varnothing, \; \forall j > 0\},$$
$$\{C : (C_j \neq \varnothing \; \forall j \leq m), \; (C_j = \varnothing, \; \forall j > m)\}.$$

Consider processing sequence $\{P\}$ by the following algorithm.

Algorithm 7.3

1. Assign $k=0$;
2. Compute w_k between P_k and R_k;
3. If $w_k = -1$, then change sequence $P_k, P_{k+1}, ..., P_n$ to sequence $C_{k+1}, ..., C_{k+m}, P_{k+1}, ..., P_n$, and assign $n=n+m$;

4. Shift sequence $C_k, C_{k+1}, ..., C_{k+m}$ to $C_{k+1}, C_{k+2}, ..., C_{k+m+1}$, and sequence $R_k, R_{k+1}, ..., R_{k+m}$ to $R_{k+1}, R_{k+2}, ..., R_{k+m+1}$;
5. While $k<n$, assign $k=k+1$ and repeat steps 2–5.

According to Section 2.3, the algorithm implements a variety of the so-called *formal T-cell*. Such a T-cell has a receptor, the type of which is stored in R_0. When the receptor is matched, the T-cell becomes *activated* and *synthesizes* a sequence of FPs $P_1, ..., P_m$, the types of which are stored in $C_1, ..., C_m$, respectively. Then the immunochip utilizes this sequence instead of P_0. Thus, moving along sequence $\{P\}$, the T-cell replaces every P_j if its type is matched with the type stored in R_0.

Designate by $S^{(k_j)}$ the type of vector that is stored in C_j. Then the function of any T-cell can be described formally by the following rule:

$$S^{(k_0)} \to S^{(k_1)}...S^{(k_m)}. \qquad (7.5)$$

Consider a correspondence between types of vectors and symbols. For example, $S^{(0)}=$ 'A', $S^{(1)}=$ 'B', $S^{(2)}=$ 'a', $S^{(3)}=$ 'b', etc. Let us take a set of $n+1$ symbols: $S^{(0)},..., S^{(n)}$. Assume that the set consists of two disjoint subsets: *nonterminals*, say $S^{(0)},..., S^{(k)}$, and *terminals* $S^{(k+1)},..., S^{(n)}$. Select one particular nonterminal (the so-called *axiom*), such as $S^{(0)}$. Now consider a set of rules (7.5) that satisfy the following conditions:

1. Any symbol of the left side is a nonterminal
2. There exists only one rule, which contains the axiom.

According to (Ginsburg, 1966), such a set of rules (7.5) is equivalent to a context-free (CF) grammar. Hence, the behavior of the set of corresponding T-cells can also be described by CF grammar. It is also worth noting that according to (Ginsburg, 1966), the class of CF grammars is the most interesting class of formal grammars for both theory and application.

7.1.3 Potential Applications for Information Security

Potential applications of the immunochip include, but are not limited to, all those considered in Chapters 5 and 6, with information security being particularly interesting (see, e.g., Sokolova et al., 2000; Skormin et al., 2001; Tarakanov, 2001). As with the biological immune system, the problem of protecting computer systems from malicious intrusions can be viewed as the problem of distinguishing "self" from dangerous "other" (or "nonself") and eliminating this "other." In the case of the computer immune system, the "nonself" may be an unauthorized user, foreign code in the form of a computer virus or worm, unanticipated code in the form of a Trojan horse, or corrupted data, etc. According to (Forrest, Hofmeyer, and Somayaji, 1997), information security could be completely specified based on the abstract representation of "self" and "nonself" as sets of bit strings, further designated as "proteins" and "peptides."

For example, a "protein" could be a sequence of viral bytes in a legitimate program, or the "signature" of a computer virus. To preserve generality, it has been proposed in (Forrest, Hofmeyer, and Somayaji, 1997) that both the protected system (self) and infectious agents (nonself) be represented as dynamically changing sets of bit strings, because in cells of the body the profile of expressed proteins (self) changes over time. Moreover, "peptide" for a computer system is defined in terms of short sequences of system calls executed by privileged processes in a networked operating system. Preliminary experiments by (Forrest, Hofmeyer, and Somayaji, 1997) on a limited testbed of intrusions and other anomalous behavior show that short sequences of system calls (currently sequences of length 6) provide a compact signature for "self" that distinguishes normal from abnormal behavior. By this analogy, proteins can be thought of as the "running code" of the body, while peptides serve as indicators of behavior.

At this point, assume that the vector X represents a set of information security indicators. It can be a bit string of a legitimate program, the signature of a computer virus, a coded sequence of system calls, statistics of current network activity, etc. Assume, then, a space $\{X\}$ of such indicators, partitioned into k subspaces (classes) $\{X\}_1,...,\{X\}_k$. This can be as simple as $k=2$, where $\{X\}_1$ is normal behavior and $\{X\}_2$ is "infection." Then, since we have a concrete vector X, the task is to determine its class $c=\{X\}_c$, where $c=1,...,k$. Thus, the problem is reduced to pattern recognition, as discussed above in Chapters 3 and 6.

This IC approach to information security allows for the use of an immunochip for on-line intruision detection, e.g., the application considered in Section 7.1.3.1. In addition, the immunochip could also be applied to some other issues of information security, like data hiding (Section 7.1.3.2) and data encryption (Section 7.1.3.3).

7.1.3.1 Intrusion Detection

For this experiment we used a fragment of a database on several types of intrusions, as shown in Table 7.1 and Table 7.2. This database utilizes a model of a typical local area network of the US Air Force (Bay, 1999).

The first column of Table 7.1 presents the conventional names of the 15 intrusion types (apache2, ..., xsnoop) and the normal behavior of the network connection (normal). The second column assigns short names to the intrusion types as listed in Table 7.2. The rest of the columns of Table 7.1 show auxiliary conditions under which the data of the intrusions had been recorded (columns 1, ..., 33 in Table 7.2).

156 Immunocomputing Systems

The database fragment in Table 7.2 presents 106 records (column #) of the several types of intrusions (column Sign). The fragment utilizes 33 characteristics (columns 1, ..., 33) of the network connection records, including lengths (number of seconds) of the connection (column 1), number of data bytes from source to destination (column 2), data bytes from destination to source (column 3), and so forth (columns 4, ..., 33).

Intrusion type	Sign	Protocol type	Service
apache2	ap2	tcp	http
buffer_overflow	b_o	tcp	telnet
guess_passwd	g_p	tcp	pop_3
Ipsweep	ips	icmp	eco_i
multihop	mul	tcp	telnet
named	nam	tcp	doman
normal	norm	udp	private
phf	phf	tcp	http
pod	pod	icmp	ecr_i
portsweep	por	tcp	private
Saint	st	tcp	private
Sendmail	se	tcp	smtp
snmpgetattack	snm	udp	private
Udpstorm	ud	udp	private
Xlock	xlo	tcp	X11
Xsnoop	xsn	tcp	X11

Table 7.1 Types of intrusions.

Consider representation of this data in a shape space by an immunochip. Consider a matrix A with dimensions 106×33 that is formed from columns 1, ..., 33 of Table 7.2. Compute the SVD of the matrix according to (3.13). Consider two FPs ({FP-1, FP-2}) as antibodies that correspond to the right singular vectors R_5, R_6. Consider also 106 FPs ({FP_1, ..., FP_{106}}) that correspond to the strings of matrix A (columns of Table 7.2). Then every string A_I that represents the number of intrusion i =1, ..., 106 can be mapped to the 2 values $\{w_1, w_2\}$ of the binding energy between FP_i and two antibodies:

$$w_1 = w(\text{FP-1}, FP_i), \quad w_2 = w(\text{FP-2}, FP_i).$$

The results are given in Table 7.3 and represented geometrically in the computer-generated Figure 7.3. Every intrusion is represented in Figure 7.3 by a point in a 2D shape space with binding energies $\{w_1, w_2\}$.

#	Sign	1	2	3	...	33
1	ap2	906	57964	0		0
11	b_o	198	2442	10661		0
14	g_p	3	36	236		0
15	g_p	0	36	236		0
16	ips	0	18	0		0
26	mul	69	331	2762		0
30	nam	707	1562	0		0
40	nom	0	105	146		0
51	phf	0	0	0		0.06
52	pod	0	1480	0		0
62	por	0	0	0		1
72	st	0	0	0		1
82	se	2	4485	1125		0
88	snm	0	105	146		0
98	ud	0	0	0		0
99	xlo	199	56124	17588		0
105	xsn	0	1256	11240		0
106	xsn	50	226	2615		0

Table 7.2 Features of intrusions.

#	Sign	w_1	w_2
1	ap2	−0.0447	−0.0184
11	b_o	0.1178	−0.0822
14	g_p	0.1126	−0.0899
15	g_p	0.1114	−0.0896
16	ips	−0.0004	0.0047
26	mul	0.1164	−0.0882
30	nam	0.1810	−0.0895
40	norm	−0.1324	−0.0189
51	phf	−0.0088	0.0033
52	pod	0.1136	−0.0903
62	por	0.1138	−0.0806
72	st	0.1158	0.0249
82	se	0.1123	−0.0895
88	snm	−0.1323	−0.0141
98	ud	0.0009	0.0036
99	xlo	0.1085	−0.0771
105	xsn	0.0981	−0.0819
106	xsn	0.1138	−0.0881

Table 7.3 Shape space coordinates of the intrusions.

158 Immunocomputing Systems

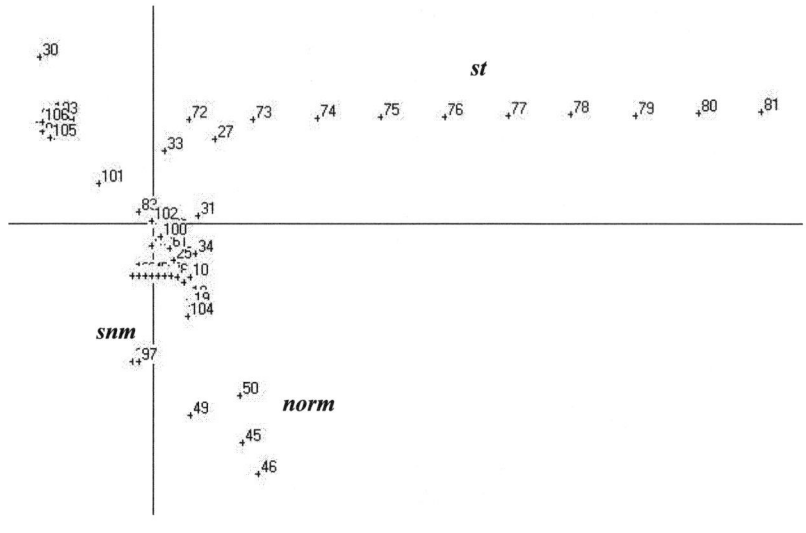

(a)

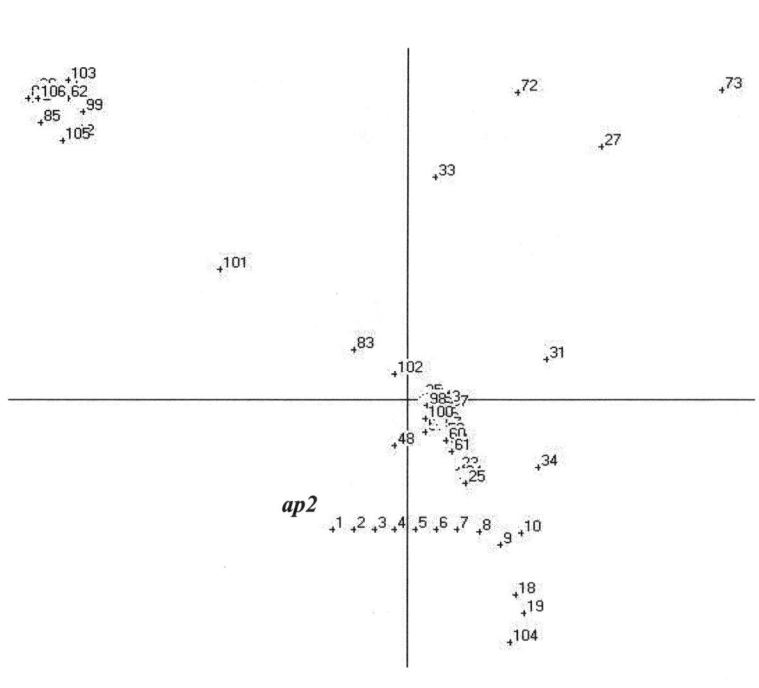

(b)

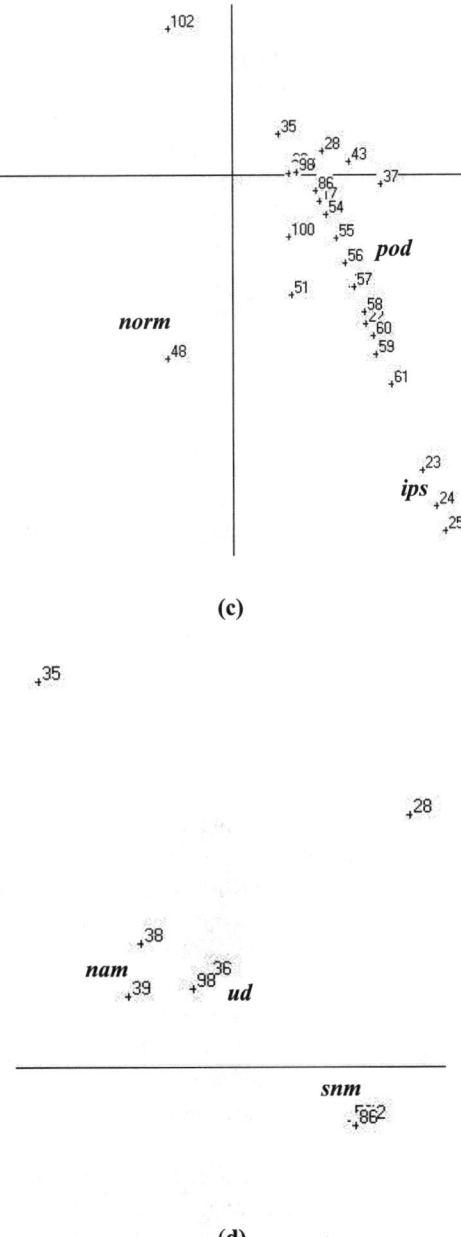

Figure 7.3 Intrusions in a shape space of the immunochip: (a), (b), (c), and (d) zoom in sequentially.

160 Immunocomputing Systems

The classification shows that this approach separates assuredly normal behavior from intrusions. Indeed, the points that correspond to normal behavior ("norm" points 45, 46, 48, 49, 50) can be separated clearly from the other points, which correspond to intrusions. Moreover, Table 7.2 and Figure 7.3 allow for reliable detection (separation) of almost all types of intrusion as relatively close points in the shape space. A more detailed analysis within the types of intrusions can be produced by zooming in with Figure 7.3.

Using this classification, the immunochip could recognize the intrusion on line in the following fashion.

Any network connection data should be recorded as a vector with 33 components:

$$X = [x_1, \ldots, x_{33}]^T.$$

The values of the components of this vector correspond to the characteristics 1 through 33 of Table 7.2.

Then two values w_1, w_2 of the binding energy should be computed according to bilinear forms:

$$w_1 = \frac{1}{s_5} X^T R_5,$$

$$w_2 = \frac{1}{s_6} X^T R_6.$$

Then the point-sample with coordinates X will be compared with every point-probe in the shape space, and the closest one will be found. This proximity could be quantified by the Euclidean norm or any other norm (see Section 3.2.1). Thus the closest point-probe determines either a normal behavior or a type of intrusion for the point-sample serving as the recognizing type of the network connection.

For example, consider row 45 of Table 7.2 as a set of characteristics of a checking network connection. Then, according to Figure 7.3a, the closest point-probe in the shape space is point 46. Since point-probe 46 corresponds to normal behavior, point-sample 45 could be recognized as the normal behavior of the checking network connection.

Another method of intrusion detection by the immunochip could also be based on supervised learning (see Section 3.2.2) without using a shape space representation. In this case, any network connection record with 33 components should be folded to a matrix M with dimensions 3×11 as follows:

$$M = \begin{bmatrix} x_1 & \cdots & x_{11} \\ x_{12} & \cdots & x_{22} \\ x_{23} & \cdots & x_{33} \end{bmatrix}.$$

In the learning mode a training set of records should be used to form matrices M_{app2}, ..., M_{xsn} for the types of intrusions and a matrix M_{norm} for the normal behavior. The SVD of each matrix results in a pair of the first left and right singular vectors $\{P_1, R_1\}$ serving as a pair of probes ("antibodies") for each learning class:

$$\{P_1, R_1\}_{app2}, \ldots, \{P_1, R_1\}_{xsn}, \{P_1, R_1\}_{norm}.$$

In the recognition mode, any checking record X should be folded to the matrix M. Then 16 values of the binding energy between the probes should be computed:

$$w_{app2} = -\{P_1^T\}_{app2} M \{R_1\}_{app2}, \ldots,$$
$$w_{xsn} = -\{P_1^T\}_{xsn} M \{R_1\}_{xsn},$$
$$w_{norm} = -\{P_1^T\}_{norm} M \{R_1\}_{norm}.$$

And finally, the intrusion type could be recognized by the immunochip by the minimal value of the binding energy:

$$w^* = \min\{w_{app2}, \ldots, w_{xsn}, w_{norm}\}.$$

Simply put, the class of the pair of probes that corresponds to the minimal value of the binding energy is the type of intrusion or normal behavior.

To improve intrusion detection by the immunochip, the two methods of recognition, supervised learning (Section 3.2.2), and unsupervised learning using the shape space (section 3.2.3), could be combined.

The approach proposed here is also able to substantially reduce the number of recorded characteristics of the network connection necessary for intrusion detection. In other words, the immunochip allows for selection of the most useful features of the network traffic by which to distinguish normal connection from attack. This could be done as follows.

Consider the coordinates R_5, R_6 of two "antibody" probes shown in Table 7.4. Every pair of these coordinates (row of Table 7.4) represents one of the 33 characteristics (first column of Table 7.4) of the network connections used in the fragment of the database of our numerical experiment.

Representation of these coordinates as points in the shape space of the IC allows for selection of the most distinctive points relative to a background of other points. Such distinctive points are shown as boldface rows 1, 15, 16, 24, 25, 26, 33 in Table 7.4. As one can see, these points have relatively large values.

characteristic #	R_5	R_6
1	**0.0883**	**0.0002**
2	0.0000	0.0000
...	...	...
14	0.0001	−0.0004
15	**0.0856**	**0.9950**
16	**−0.0104**	**0.0157**
17	0.0003	0.0015
...	...	...
23	−0.0001	0.0002
24	**0.4163**	**−0.0777**
25	**−0.9008**	**0.0585**
26	**−0.0037**	**0.0013**
27	0.0011	0.0008
...	...	...
32	0.0007	0.0014
33	**0.0023**	**0.0052**

Table 7.4 Coordinates of "antibody" probes.

Another numerical experiment was staged using this reduced set of seven characteristics. A matrix M with dimensions 106×7 was formed with the input data of Table 7.2 (rows 1–106 and columns 1, 15, 16, 24, 25, 26, 33). The SVD of this matrix was computed, and so forth. Finally, the shape space representation of the data was computed. Remarkably, the reduced set of 7 selected characteristics gives almost the same results as the full set of 33 characteristics in Figure 7.3. The reduced set also separates assuredly normal behavior from intrusions and reliably detects almost all types of intrusions as relatively close points in the shape space.

In other words, the IC approach has reduced the amount of necessary initial data of network connection records by almost five times, from 33 characteristics to 7, without losing intrusion detection capabilities. Therefore, the proposed immunochip is able to select, on line, the most useful traffic features by which to distinguish a normal connection from an attack. As an example, the description of these features for the data of the computation experiments are given in Table 7.5.

feature #	feature name	Description
1	Duration	length (number of seconds) of the connection
15	Count	number of connections to the same host as the current connection in the past two seconds
16	srv_count	number of connections to the same service as the current connection in the past two seconds
24	dst_host_count	number of connections to the same destination host as the current connection in the past two seconds
25	dst_host_srv_count	number of connections to the same destination host and the same service as the current connection in the past two seconds
26	dst_host_same_srv_rate	% of connections to the same destination host and the same service as the current connection in the past two seconds
33	dst_host_srv_rerror_rate	% of connections that have errors

Table 7.5 Most useful traffic features for attack detection.

7.1.3.2 Data Hiding

According to (Bender et al., 1996), data hiding, a form of *steganography*, embeds data into digital media for the purposes of identification, annotation, and copyright with a minimum amount of degradation to the "host" signal; i.e., the embedded data should be invisible and inaudible to a human observer. Note that data hiding, while similar to compression, is distinct from encryption. Its goal is not to restrict or regulate access to the host signal, but to ensure that embedded data remain inviolate and recoverable.

As an example of data hiding by the immunochip, consider that matrix A represents an initial data array. It could be an image, a folded audio signal, etc. Consider the SVD of the matrix in the form of (2.9). Let us add to this sum an item in the form $s_{r+1}L_{r+1}R^T_{r+1}$, where r is a rank of the matrix, L_{r+1}, R_{r+1} are unit vectors, s_r is a minimal singular value of the matrix, and $s_{r+1} < s_r$. According to (Horn and Johnson, 1986), such an addition perturbs the matrix only slightly. Although such a perturbation is invisible or inaudible to a human observer, the presence of the "hidden" addition can surely be detected in the shape space of the IC. In this case the immunochip functions like the natural immune system, which verifies identity by the presence of peptides, or protein fragments.

7.1.3.3 Data Encryption

Next consider data encryption. In modern *cryptography*, encrypting of information is based on a widely known algorithm and a secret number or string, called a key. The key is used as a parameter to the algorithm to encrypt and decrypt the data. Decryption with the key is simple, but without the key it is very difficult and in some cases nearly impossible. Therefore, the "fundamental rule of cryptography" is that both the sending and receiving sides know the method of encryption (Tannenbaum, 1996).

As an example of encryption by the immunochip, consider BB-networks. Specifically, in the network $1DBB(10,2)$, for any type $i=0,...,9$ there exist populations of the type

$$S^{(i+2)} S^{(i)} S^{(i-2)} S^{(i)},$$

which, according to (Tarakanov, 1999a), is cyclic with period 4. For example,

$$1979 \to 187800 \to 1770991 \to 17980 \to 1979 \to \cdots,$$

where type $S^{(i)}$ is denoted by only one number i. Let such numbers code letters according to Table 7.6:

B-cell type	0	1	2	3	4	5	6	7	8	9
Letter	c	l	x	x	x	x	x	e	p	o

Table 7.6 Coding of letters by types of B-cells.

Then network dynamics can be presented as

$$\textit{\textbf{loeo}} \to \textit{\textbf{lpepcc}} \to \underline{\textit{\textbf{leecool}}} \to \textit{\textbf{leopc}} \to \textit{\textbf{loeo}} \to \cdots.$$

Let the numbers 10 and 2 serve as a key that defines the network $1DBB(10,2)$. Then the string ***loeo*** (1979) could encrypt the sentence <u>***leecool***</u> (1770991). With knowledge of the key, the sentence can be decrypted, for instance, as the string of the maximal length generated by the network $BB(10,2)$ from the given string <u>***loeo***</u> (1979). Although the example seems rather simple, it nevertheless demonstrates the principal possibility of using the immunochip in cryptography.

7.2 An Immunochip Emulator

Within the IC approach, we are also developing a software emulator of the immunochip. The goal is twofold:

1. To raise the quality and compliance of IC models by implementing and studying all of them on the emulator;
2. To improve the emulator itself using several models.

The emulator is being developed utilizing the following standard software:

- MS Windows 2000 Operating System;
- MS Visual C++ 6.0 Developer Studio;
- OpenGL tools.

In this section we present preliminary versions of the emulator:

1. Version 1.0: Excitable Microarrays;
2. Version 1.1: 3D Excitable Microarrays;
3. Version 2.0: Self-Assembly of FP;
4. Version 3.0: 2D FIN;
5. Version 3.1: 3D FIN;
6. Version 3.2: Immune Response in 2D FIN;
7. Version 4.0: Orbital Hodograph.

7.2.1 Version 1.0: Excitable Microarrays

The goal of this version is to probe the emulator as a universal computation medium (Adamatzky, 1997).

The CHIP consists of 4 microarrays or layers, as shown in Figure 7.4.
Every ARRAY consists of cells (currently 100×100 cells).
Every CELL can be in one of only 3 states:

- 0 rest (blue);
- 1 refracted (green);
- 2 excited (red).

Each cell interacts only with its nearest neighbors:

- 8 horizontal (of the same array);
- 3 vertical (of the other arrays).

Rules of interaction:

0. *If* the cell is at REST
   and the sum of excited neighbors
      lies in some specified interval {th_1, th_2},
   then the cell becomes EXCITED;
1. *If* the cell is REFRACTED, *then* it is at REST;
2. *If* the cell is EXCITED, *then* it becomes REFRACTED.

Note: All cells change their states simultaneously.

Examples of generating modes dependent on excitation thresholds {th_1, th_2}:

- {0, 1}: morphogenesis (formation of stable patterns);
- {1, 8}: chemical waves;
- {2, 2}: solitons.

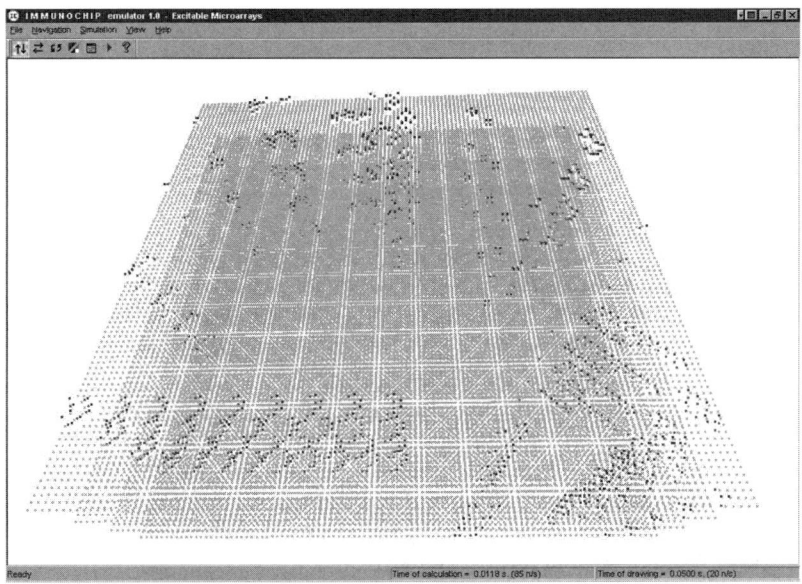

Figure 7.4 Two-dimensional solitons.

7.2.2 Version 1.1: 3D Excitable Microarrays

The goal of this version is to probe the emulator as a tool for 3D modeling of dynamic processes.

The CHIP consists of 3D microarrays, as shown in Figure 7.5.

Every ARRAY consists of cells (the emulator currently includes 50×50×50 cells).

Every cell interacts only with its nearest neighbors:

- 8 horizontal (on the same array);
- 18 vertical (on the others arrays).

The states of the cells and the rules of interaction are as in Version 1.0. Examples of generating modes dependent on {th_1, th_2}:

- {0, 1}: morphogenesis;
- {4, 8}: steric waves;
- {3, 3}: steric solitons.

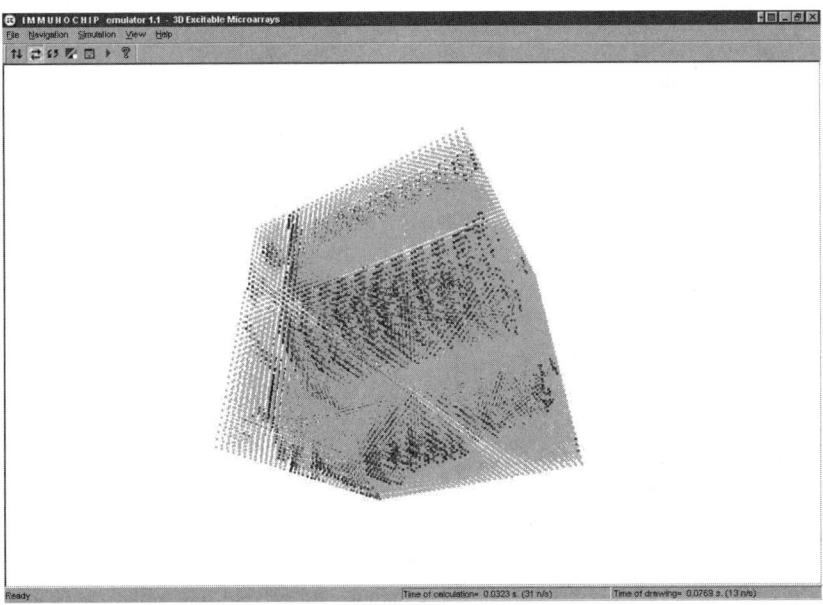

Figure 7.5 Three-dimensional waves.

7.2.3 Version 2.0: Self-Assembly of FP

The goal of this version is to probe the emulator as a tool for fast 3D modeling and visualizing the process of self-assembly of the FP.

The CHIP consists of 2 microarrays (layers, currently 180×180 cells each):

168 Immunocomputing Systems

1. The energy layer (left hand screen in Figure 7.6) computes and shows the surface of the free energy of the FP, also called a "Ramachandran diagram" in biophysics (Cantor and Schimmel, 1980a);
2. The angles layer (not shown on the screen) stores a table of torsion angles for fast minimizing of the free energy of the FP during self-assembly.

To make computations visual, we also represent (Figure 7.6, right-hand screen) the current torsion angles of the FP as the angles of the real-world skeleton of the protein. Color-coding of the links of the protein's skeleton is given in Table 7.7.

Link	Color
N – H	Cyan
N – C^a	White
C^a – H	Cyan
C^a – C	White
C – O	Blue
C^a – Amino Acid Residue	Red
C – N (peptide's bond)	Yellow

Table 7.7 Color-coding of the links of the protein's skeleton.

This version allows the user to probe a self-assembly of the FP dependent on the number of links of the FP and the values of the controls of the quadratic form (2.3).

7.2.4 Version 3.0: 2D FIN

The goal of this version is to apply the emulator as a tool with which to model an FIN.

The CHIP consists of 2 microarrays (layers):

1. B-cells layer (left-hand screen in Figure 7.7)
2. Energy layer (not shown on the screen)

Each type of B-cell is coded by a definite color.

The current number of B-cells of each type is shown with colored bars (right-hand screen in Figure 7.7).

A user can assign two main parameters for the $2DBB(n, n_h)$ emulator: n for the types of B-cells and n_h for the threshold of binding. Examples of the main modes are shown in Table 7.8.

7.2 An Immunochip Emulator   169

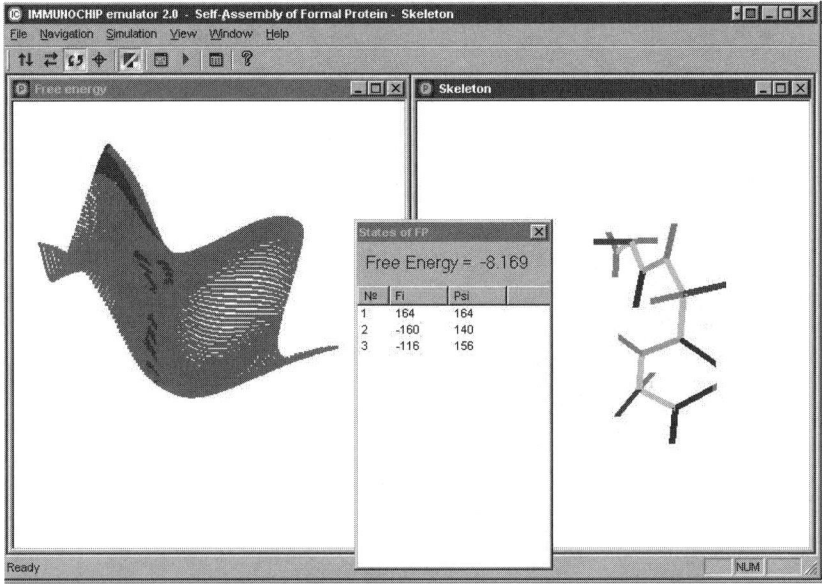

Figure 7.6 FP folding.

Mode	(n, n_h)	Model
Stable	(5, 2), (7, 3), (9, 4), (11, 5)	Dynamic immune memory
Disorder	(6, 2), (7, 2), (8, 3)	Immune disease
Quasi-chaotic (fractal)	(16, 2), (50, 2)	Immunodeficiency
Overflow (texture)	(any, 1)	Allergy

Table 7.8 Modes of the immunochip emluator.

At each time step the emulator also computes and shows the following parameters of 2D_FIN:

- number of cell types;
- threshold of binding;
- total number of cells in FIN;
- FIN size $i \times j$;
- table of cell types with current number of cells.

170 Immunocomputing Systems

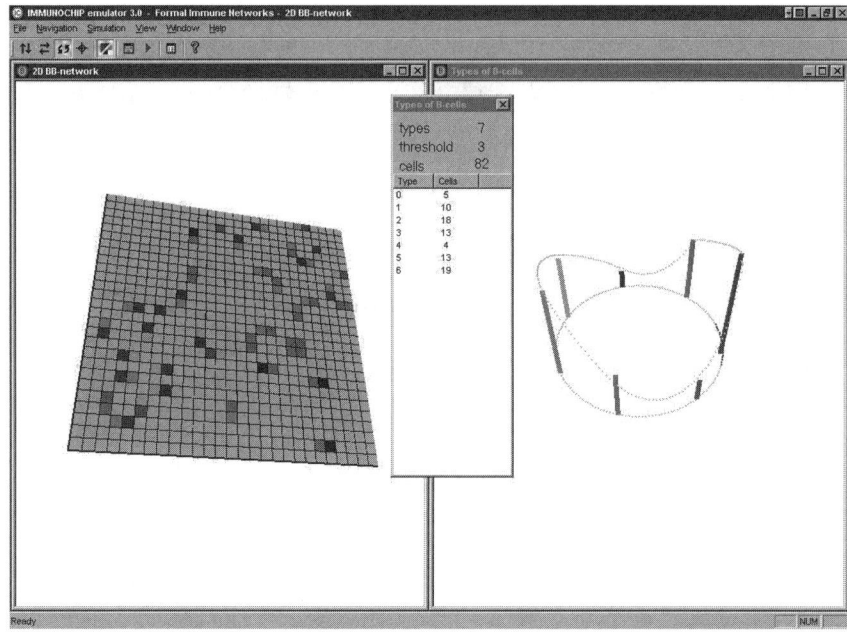

Figure 7.7 Two-dimensional FIN

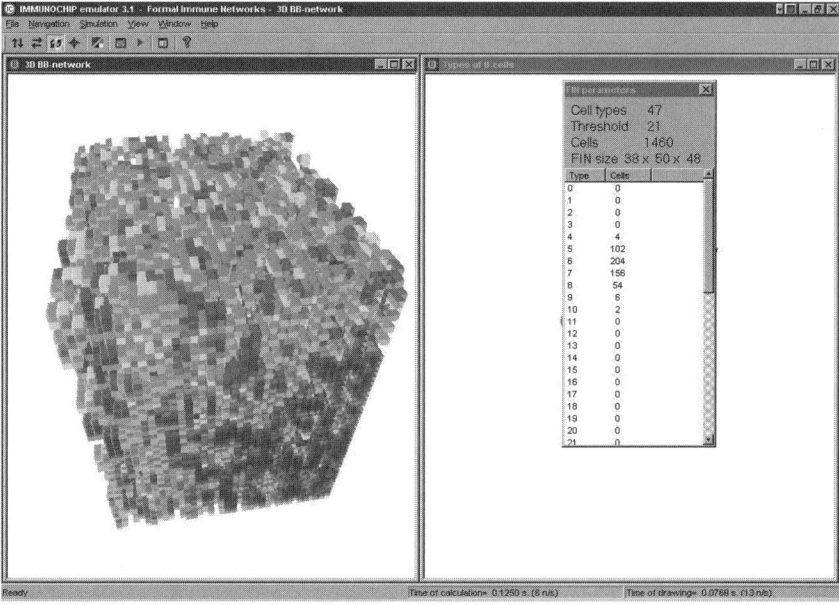

Figure 7.8 Three-dimensional FIN.

7.2.5 Version 3.1: 3D FIN

The CHIP consists of 3D microarrays (currently the emulator includes 50×50×50 cells):

1. B-cells layers (left-hand screen in Figure 7.8)
2. Energy layers (not shown on the screen)

All other features of the emulator correspond to Version 3.0, excluding FIN size, which is now 3D ($i{\times}j{\times}k$), and 3D modes of the emulator defined by values of (n, n_h) other than those in Table 7.8.

7.2.6 Version 3.2: Immune Response in 2D FIN

The goal of this version is to apply the emulator as a tool for modeling of the immune response in FIN.

CHIP consists of 3 microarrays (layers):

1. B-cells layer (upper layer in the left-hand screen in Figure 7.9)
2. Antigens layer (lower layer in the left-hand screen in Figure 7.9)
3. Energy layer (not shown on the screen)

Each type of B-cell and antigen is coded by definite color.

The user can assign the following parameters of the $2DAB(n, n_h)$ emulator: types of B-cells (n), threshold of binding (n_h), mutation rate as a number {0, 1, 2, 3, 4} of the mutated daughter cells for each B-cell, presence or absence of the antigens {Yes, No}, and if Yes then the type of the antigens {0, 1, ..., $n-1$}.

For each time step the emulator computes and shows the following parameters of 2D FIN:

- number of B-cells and antigens types;
- threshold of binding;
- mutation rate;
- type of antigens;
- total number of B-cells in FIN;
- FIN size $i \times j$;
- table of cell types with current number of B-cells.

An example of the immune response in a two-dimensional AB network is shown in Figure 7.9. In this example, FIN begins to generate the increased level of antigen binding B-cells of type 2 (see southwest corner of the B-cells layer).

172 Immunocomputing Systems

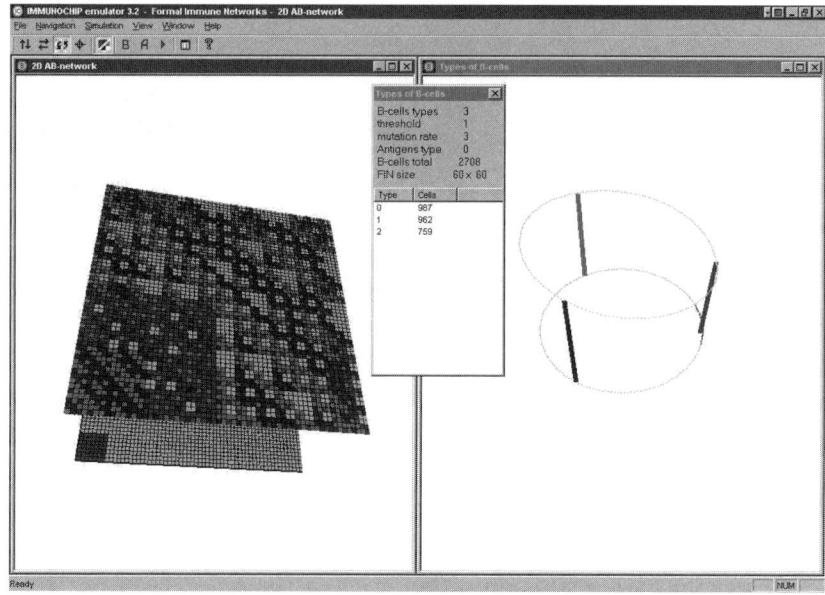

Figure 7.9 Immune response in 2DAB FIN.

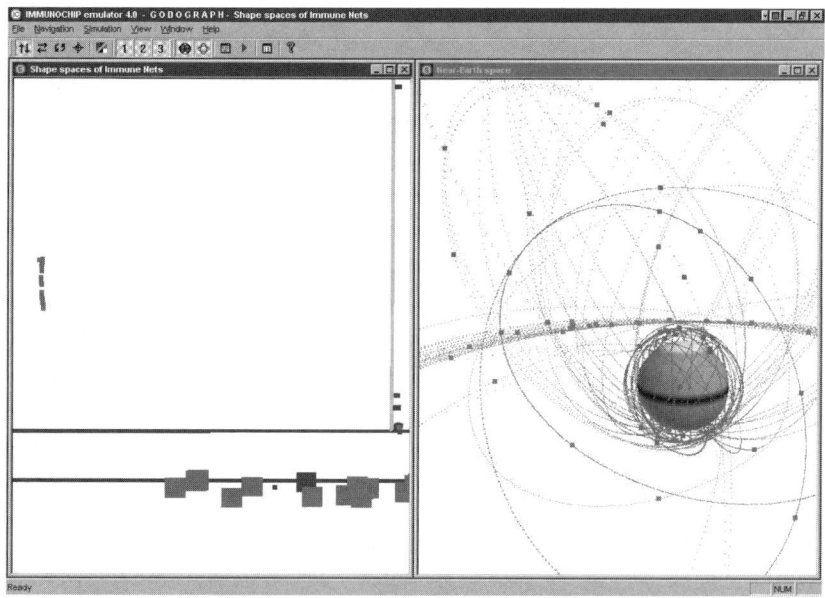

Figure 7.10 On-line detection of dangerous situations in near-Earth space.

7.2.7 Version 4.0: Orbital Hodograph

This chip is intended to compute immune and shape space representation (left-hand screen in Figure 7.10) of the global ballistic situation in near-Earth space (right-hand screen in Figure 7.10) using the theory of an orbital hodograph (see Section 6.1).

The input data of the emulator are parameters of the orbits (see, e.g., Table 6.1) from an external file.

Currently, the emulator consists of three shape spaces (layers), which compute immune interactions between the following pairs:

1. Orbit–Earth
2. Orbit–Orbit
3. Satellite–Satellite

Each such pair is shown as a colored bar on the left hand screen.
This immune simulation can be extended to include:

4. Satellite–Point
5. Satellite–Airplane

Dangerous pairs are filtered by the horizontal axis on the left-hand screen (shape spaces of the emulator).

The emulator also shows other parameters of the simulation at the user's request, including:

- total number of satellites;
- current time;
- simulation time;
- a table of parameters of the orbits.

7.3 The Biochip

The development and application of arrays of immobilized biological compounds (biological microchips, or biochips) have become a significant trend in modern biology, biotechnology, and medicine (Cheng and Kricka, 2001). This technology brings together computer chips with biomedical assays (or tests) and laser-based detectors (see, e.g., Swanson et al., 2000). Biochips make possible the manipulation of proteins and other biomolecules to perform immunoassays for high-throughput, precise, and economical diagnostics, and even for computer-controlled extensions of the natural immune system in the near future.

The main advantage of biochips over conventional analytical devices is the possibility of massive parallel analysis. Biochips are smaller than conventional testing systems and highly economical in their use of specimens and reagents, whereas traditional methods often include long incubations and extensive manual steps. Major progress has been achieved in the manufacture and application of DNA microchips, but DNA is not the only biological entity that can be arrayed or spotted onto the surface of these biochips. It is also possible to configure arrays of proteins, but it is a very difficult task (see, e.g., Protein chip challenges, 2000). Proteins cannot be immobilized on the same types of materials as DNA, for they are extremely sensitive to the physical and chemical properties of the particular substrate. Some substrates will work for one class of proteins but not another, making it difficult to come up with a so-called generic, solid-phase material. The bound proteins must remain active and must be able to select the proteins of interest from the experimental sample (complex samples such as whole blood or tissue). The proteins cannot be amplified like nucleic acids, so improvements must come from other areas, which may also include new imaging technologies or labels for detecting reactions. The development of protein biochip fabrication and its applications is just beginning. There have been very few reports in which the antigen–antibody reaction was successfully emulated in a biochip. Several types of immunoassays were performed in which either the antigen or the antibody was immobilized onto or within different substrates and detected directly with labeled antibodies or antigens.

Therefore, as a natural outgrowth of the IC approach, we propose to create a preactivated biochip prototype ready to array DNAs and proteins (antibodies, antigens, receptor proteins, etc.). This would constitute the basis for a new generation of clinical diagnostic analyses, as well as for a biomolecular computer. Two applications of the biochip prototype for medical diagnostics of C-reactive protein and microalbuminuria are also included in the project.

Our biochip will include the following main components:

- *Biochip Matrix*: thin, monolithic, macroporous membranes with an array of microwells;
- *Protein Streptavidin* microarraying: a molecular adapter attached to the surface of the microwells and facilitating the application of DNAs and proteins samples;
- *Biochip Comtroller*: an associated liquid delivery and scanning detector system with a computer-aided procedure for user-friendly analysis results of diagnostics at each microwell.

We propose to develop three products: the basic prototype and two applications.

The first product, a preactivated biochip prototype, will allow a user to immobilize in the microwells a wide range of molecular probes and to detect their binding with the samples. Thus, the whole system could be applied for many clinical diagnostic assays.

The second product, the biochip-based diagnostic assay for the detection of C-reactive protein, will allow a user to detect tissue destruction, necrosis, and inflammation in the early stages of diseases, including atherosclerosis and ischemia.

The third product, the biochip-based diagnostic assay for the detection of microalbuminuria, will allow a user to detect the subset of diabetic patients at risk of developing renal failure and early cardiovascular morbidity and mortality.

The proposed biochip will allow users to test hundreds of samples simultaneously and will accelerate diagnostic time by a factor of 3 to 8. The biochip will be much smaller than conventional testing systems, highly economical in the use of specimens and reagents, ultrasensitive to a very small amount of sample, and the cost of one "microassay" fulfilled by the biochip prototype will be comparable to the cost of standard immunological diagnostic assays.

The expected users of the first product are manufacturers of diagnostic assays, as well as research laboratories. The expected users of the second and third products are clinical diagnostic laboratories and diabetic centers.

The biochip-based microalbuminuria detection assay and C-reactive protein detection assay will provide screening checkup programs for microalbuminuria and persistent proteinuria (to detect the subset of diabetic patients at risk of developing renal failure and early cardiovascular morbidity and mortality) as well as checkup and screening programs for patients with atherosclerosis, cardiovascular diseases, and individuals at increased risk of developing myocardial infarction (according to definite, age-matched risk groups). Those patients found to be at risk should be regularly monitored. The possibility of massive parallel analysis is the main advantage of biological microchips over conventional analytical devices.

7.3.1 The Biochip Matrix

According to (Protein chip challenges, 2000), biochips do not necessarily have to be silicon wafers. They can be made of aluminum, for instance, or even glass. And they do not have to be wafers; slides are also used. Also, some companies are developing microfluidic chips, on which it is possible to perform lab-type assays in situ. And most importantly, DNA is not the only biological entity that can be arrayed or spotted onto the surface of these "chips": it is also possible, but difficult, to configure arrays of proteins.

So, the important step in biochip production is the development of a support matrix. The chosen support must have a number of desirable properties (high binding capacity, low nonspecific binding of biomolecules, and exceptional chemical stability). We propose to use as a biochip matrix thin monolithic macroporous membranes, also called convective interaction media (CIM), made of a special modified cross-linked biopolymer. They are designed for the isolation and purification of proteins, peptides, oligo- and polynucleotides, as biosensors, and for usage in diagnostic kits.

CIM supports exhibit low nonspecific binding of biomolecules and offer exceptional chemical stability. CIM membranes are designed for laboratory and industrial isolation and purification of peptides, proteins, oligonucleotides, and polynucleotides. CIM are perspective chromatographic and bioconversion supports based on a highly cross-linked porous monolithic polymer. They are available in different forms and chemistries, making them suitable for nonexchange, hydrophobic interaction and affinity chromatography, as well as bioconversion purposes. The porous polymer has a mean pore size of 800 nm, while the size may be varied and optimized. The polymer also has a carefully balanced hydrophobic–hydrophilic matrix that allows both simple attachment of ligands—which include biological molecules and their fragments—onto the surface of pores within the monolith and preparation of an efficient medium for "protein arraying." Immobilization of protein ligands to CIM may take place via the epoxy groups found naturally in the polymer structure of CIM material and the amino groups of proteins.

However, the main advantage of using CIM as a biochip matrix could be its compatibility with the biological tissues. Thus, in perspective, such biochips could be used also as implants.

7.3.2 The Molecular Adapter

The protein candidate to be attached to the biochip matrix is streptavidin (SA). Streptavidin, a tetrameric protein produced by *Streptomyces avidini*, is known for its extraordinarily high affinity with biotin (Bi). Bi is a vitamin (vitamin H) synthesized by plants, most bacteria, and some fungi. Nowadays, Bi is widely used as a reagent in biotechnology, biochemistry, and immunology. SA is one of the most stable proteins known; it maintains its functional structure at high temperatures, extremes of pH, in the presence of high concentrations of denaturants and organic solvents, and it is exceptionally stable against proteolysis. The hydrophobicity of SA yields a very strong binding to the solid phase.

The interaction between SA and Bi is one of the tightest noncovalent interactions known between proteins and their ligands (Stayton et al., 1999). So the SA binding of biotinylated proteins is also very tight and stable, and the binding becomes independent of the nature of the protein (its hydrophobicity or hydrophilibity, and its charge differences: isoelectric point, accessibility of active functional sites after coating processes, etc.).

Biotin can be easily and effectively attached to different molecules, including proteins and nucleic acids, without destroying their biological activity. Biotinylation of proteins is a very gentle method, so inactivation of the protein by coupling procedures can be reduced considerably. Antibodies and antigens bound to the matrix via SA/Bi retain their immunologic reactivity to a much higher degree than those directly attached to the solid phase. Thus, interaction between antibody and antigen is enhanced and more stable.

Immobilization of DNA molecules to supports via an SA/Bi link is particularly useful, since the bond can be established rapidly, and it remains stable under extreme physical conditions, including temperatures and pH that denature DNA molecules into single strands. So SA is not only one of the most stable proteins, but it is a universal reagent for application in diagnostic tests based on protein and DNA detection.

The high-affinity recognition of biotin (Bi) by streptavidin (SA) has made this protein one of the most useful tools in biotechnology. It is widely used in diagnostic assays and in a variety of other applications. These methods have collectively become known as streptavidin–biotin technology. Biotin can easily and effectively be attached to different molecules, without destroying their biological activity. The exceptional stability of the SA/Bi complex explains the popularity of this system.

The SA-bound CIM support can be used to immobilize numerous types of molecules onto a substrate in addressable patterns. For example, biomolecules such as nucleic acid molecules or fragments thereof, isolated proteins or fragments thereof, carbohydrates, protein binding (receptor) proteins, etc., can be immobilized onto a support via SA/Bi interaction.

In the SA-bound CIM support, SA serves as "molecular adaptor" in solving many of the problems listed above. The standardization of the coating processes and the highly improved and enlarged capture matrix (via SA/Bi interaction) have the following consequences: Test sensitivity can be increased up to ten times, and incubation times can be reduced from hours to minutes. The use of the SA/Bi link to immobilize and manipulate nucleic acids can simplify many aspects of gene analysis. SA/Bi technology provides ultrasensitive tests in the clinical diagnostic field for detection of both normal and elevated target levels. Thus the technology is extremely useful for discovering very early events of disease as well as very small amounts of targets. This feature is highly important for medicine, because the earlier the disease can be discovered, the higher the chances for successful treatment.

Combining SA/Bi technology and biochip technology gives rise to the following innovations: the development of protein biochips combining universality in application, ultrasensitivity, miniaturization, short time, and the ability to analyze thousands of samples simultaneously.

7.3.3 The Biochip Controller

The function of the Biochip Controlling System (BCS) will be twofold:

1. as a liquid delivery system of the biochip;

2. as a scan reader of biochip reactions.

The BCS will contain:

- a high-precision, two-coordinate pointing mechanism;
- an auto-manual or an automated dispense module;
- a precision optical electronic detection module.

The two-coordinate pointing mechanism and the dispense module will be controlled by means of a PC-compatible computer. We intend to use two types of dispense modules available on the market: auto-manual and automated. Accordingly, our BCS with an auto-manual dispense module could be used in medical laboratories, while that with the automated dispense module could be used for industrial applications.

We intend to use a modification of the available electronic pipettor as the auto-manual dispense head. This pipettor is a multifunction instrument. It provides high accuracy for the reagent's loading, diluting, mixing, and dispensing by means of an embedded microprocessor. The electronic pipettor will be moved manually to the pipette washing submodule and the reagent's loading submodule. These submodules will be arranged separately.

The automated dispense module unites the dispense head, washing submodule, and loading submodule in one system. All operations are performed by this dispense module automatically.

The functioning of the BCS includes two steps. First, the biochip will be placed on the horizontal surface of the two-coordinate stage under the dispense head. The reagents will be deposited on the biochip in a scanning-type motion with respect to the dispense head. Second, the biochip will be moved by means of the two-coordinate stage to the detection module and scanned by the detection head.

The digital or analog signals from the detection module will be transmitted to the PC computer for processing through a standard serial port or a special PCI card.

It worth noting that in conventional immunoassays and in the immunoassays of biochip types, the following reagents are usually used for the detection of antigen–antibody interaction: fluorescent dyes, antibodies, or Fc-receptor proteins coupled with enzymes that catalyze the enzyme–substrate reaction with colored products. All of these lead to the development of complicated and expensive biochip readers of the spectrophotometric type.

However, an essential feature of the biochip proposed here is a new type of detection of bindings by covalent coupling of target molecules with activated and modified carbon particles (Plaksin, Raev, and Gromakovskaya, 1997). This method is an alternative to antibodies conjugated with enzymes (horseradish peroxidase or alkaline phosphatase) as labeling reagents, and the method excludes an enzymatic substrate reaction or application of fluorescent dyes. The detection reagent developed here has several advantages, including ease of use and interpretaion, shortened assay time due to the elimination of the enzymatic-substrate reaction stage, high sensitivity in comparison to similar methods using standard horseradish peroxidase- or alkaline phosphatase-labeled conjugates, stability of the detection reagent, and nontoxicity.

This innovation also simplifies the biochip reader and makes it much cheaper, especially in relation to the use of fluorescent dyes and their appropriate equipment. Our BCS will use a combined method of analysis of biochip reactions based on photometric and imaging detection procedures. This method will be applied for the analysis of optical and spatial parameters of reaction locations on the biochip. Each location will include the carbon particles labeled reagent system (CP system). The photometric procedure will provide reflection measuring of the optical density of the CP system. This procedure will be basic for the analysis of the reactions by the BCS. The imaging procedure will detect the spatial parameters of the CP system. This procedure will be additional for the BCS. The combination of these two procedures should provide certain detection of the results of the biochip reactions.

Software for user-friendly analysis of the results of the diagnostics will also be developed as a version of the immunochip emulator (see, e.g., Figure 7.11).

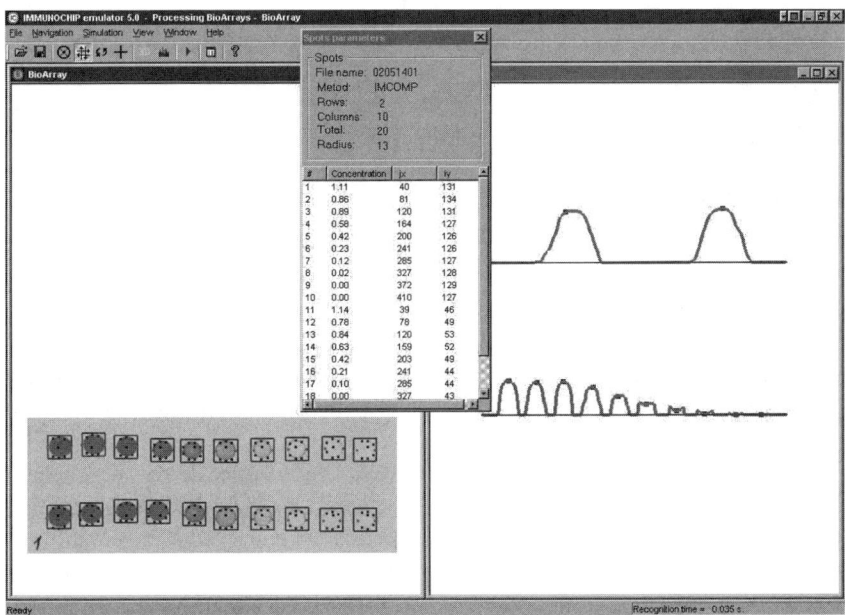

Figure 7.11 Processed results of the reactions (nigrescence) in each location of the biomembrane.

7.3.4 Biochip Applications

Consider two important applications of the biochip to early diagnostics of C-reactive protein (CRP) and microalbuminuria.

Application-1: C-Reactive Protein

CRP is a reliable biochemical marker suitable for detection of tissue damage, necrosis, and inflammation. CRP is an essential human acute-phase reactant produced in the liver in response to systemic stimuli. CRP values in acute bacterial infections have been appreciated for 70 years.

In recent years it has been established that inflammation is a key mechanism in the pathogenesis of atherosclerosis and in coronary artery disease progression. Atherosclerosis and other pathologies involving inflammation are associated with increased levels of cytokines, which in turn raise acute-phase protein levels in blood (acute inflammation markers), CRP among them. It has been shown that concentration of CRP is higher in individuals at increased risk of developing cardiac events. This is true both in apparently healthy men and women and in ischaemic heart disease patients. That is why elevated plasma levels of CRP (and higher base-line plasma CRP concentrations) in patients with unstable coronary syndromes and in population studies are considered to be predictive of future adverse events, including cardiac death and myocardial infarction, implicating inflammation in pathogenesis. The number of patients with atherosclerosis and cardiovascular diseases increases every year and requires massive screening of high-risk groups of patients defined by conditions and age.

Taking into consideration these facts, an ultrasensitive CRP assay carried out in shortened time, with an exceptionally high level of sensitivity that has not been compromised by the size of its biochip, is a matter of current interest.

Application 2: Microalbuminuria

Today, about 100 million diabetic patients are registered throughout the world. The number of newly registered cases of diabetic disease increases 7 to 12% every year and increases twice every 6 to 7 years.

Diabetic nephropathy is a major cause of premature death in diabetic patients, largely from uraemia and cardiovascular disease. Diabetic nephropathy develops in about 30% of people with insulin-dependent diabetes mellitus. Diabetic renal disease is a multistage condition that requires several years to become clinically overt. At the onset of the disease, there are usually changes in renal function, such as glomerular hyperfiltration, increased renal blood flow, and hypertrophy of the kidney. At an early stage, most of these changes can be reversed, but they persist in many patients and may have prognostic significance for the later development of clinical nephropathy. These early changes seem to occur both in insulin- and non-insulin-dependent diabetic patients.

The first sign of diabetic nephropathy is a persistent increase in the albumin excretion rate to 20–200 μg/min (30–300 mg/24h). This phenomenon is called microalbuminuria (MAU) and may be detected after one year of diabetes in insulin-dependent diabetic patients of postpubertal age and at diagnosis in non-insulin-dependent diabetes. MAU indicates incipient nephropathy. MAU in both types of diabetes is predictive of persistent proteinuria and early death from cardiovascular disease. Moreover, MAU is associated with a higher prevalence of retinopathy (particularly in IDDM), peripheral vascular disease, and neuropathy. MAU that develops in the first few months of a diabetic pregnancy may signal an increased risk of preeclampsia.

We propose to use the standard method for detection of albumin concentration in urine, based on immunoassay, but we will replace the anti-albumin antibodies that are traditionally used in the immunoassays with recombinant albumin binding protein. Recombinant albumin binding protein highly specifically binds only albumin and no other proteins.

The affinity of such binding is very high, and the binding itself has been better standardized in comparison with polyclonal and monoclonal antibodies, the polyclonal antibodies being highly heterogenic and monoclonal antibodies possessing less affinity.

This biochip-based MAU detection assay will provide a screening checkup program for MAU and persistent proteinuria to detect the subset of diabetic patients at risk of developing renal failure and early cardiovascular morbidity and mortality. Those patients found to be at risk should be regularly monitored, and the possibility of massive parallel analysis is the main advantage of biological microchips over conventional analytical devices.

7.3.5 From Biochip to Biocomputer

In perspective, two steps could transform the proposed biochip to a biomolecular computer: (1) an immunochip-based controller, including (2) controlling of biomolecules in the microwells of the biochip.

The first step is needed to modify a PC-compatible computer by the addition of an immunochip. At this step the function of the immunochip is twofold: (1a) control of the liquid delivery system of the biochip and (1b) control of the biochip reader.

The first step—the immunochip-based biochip controller—is necessary for control reactions in the microwells of the biochip. For example, the immunochip could automate feeding the biochip with reagents and samples, removing intermediate products, changing probes in the process of training of the biochip, etc. Although such functions seem complicated, it is worth noting that they are currently under development for some microfluidic biochips (e.g., see Protein chip challenges, 2000). Also at this step the immunochip should provide a surveillance of the biochip surface, including image processing from the biochip reader and recognition of the results of the reactions in the microwells.

The second step presents a solution to the key problem of the biocomputer: providing control of biomolecules in the microwells of the biochip by a molecular-electronic impact computed and performed by the immunochip. Simply put, the biocomputer could form biomolecules with needed properties at appropriate locations on the biochip. If the problem is solved, the next step could be secretion of necessary biomolecules at appropriate times. For example, in this way the implanted biocomputer could control and correct natural immunity.

Therefore, the above two steps would allow us to obtain a full-value biocomputer, where natural biomolecules (proteins and DNAs) of the biochip collaborate with the silicon schemes of the immunochip.

It is worth noting that the choice of the CRP for one of the variants of the biochip is not accidental. The functions of this protein are close to those of cytokines: special proteins secreted by immune cells to control immune response. It is known that violation in synthesis and secretion of the cytokines could cause several violations of immunity. Therefore, the development of the biochip for detection of such proteins is also a step by the biocomputer for evaluation and control of the cytokine system in general. So, the development of the biocomputer to control the cytokine complex in model biological microsystems (in vitro) as a fragment of the computer-controlled immune system seems quite realistic and well-timed.

Conclusion

The authors realize that the IC approach presented is this book is still very new. Unavoidably, it contains several gaps that will need to clarified and resolved in the future. However, the following three features unquestionably make this concept extremely promising:

1. The highly appropriate and efficient biological prototype of the described immune networks;
2. The rigorous mathematical basis of FIN;
3. The feasibility of hardware implementation in a special immunochip.

These realities could raise the IC as well as its principal applications (e.g., to information security) to a level of reliability, flexibility, and operating speed that is far beyond any conventional computer and that has never been achieved by neurocomputing.

Proliferation of the IC and its implementation in an immunocomputer would result in the following innovations in computing technology:

- The main drawbacks of neurocomputers (spurious patterns, low capacity in relation to the size of neural network, difficulty with location of errors, etc.) would be overcome. These drawbacks block the wide application of neurocomputers in fields where errors cost too much (e.g., control and navigation of spacecraft, aircraft, ships, submarines, security systems, intensive care medicine);
- IC provides an effective simulation of the natural immune system. Even the simplest versions of FIN effectively simulate important properties of immune response and immune memory that could be very useful in future medical research into the nature of the body's reaction to deadly diseases such as AIDS;
- Enabling diagnostic systems to process huge amounts of data in real time, thus ensuring dependable detection and prediction of failures and critical situations in spaceships, aircraft, nuclear power plants, and ecological systems;
- Advancing the field of information security by providing the basis for the development of self-learning security systems that can resist

unknown invaders (viruses, unauthorized users) and software/hardware implementation of security systems;
- Improving the reliability and behavior flexibility of computer-controlled mobile objects (robots, etc.) in unpredictable situations;
- Enhancement of data mining applications for the detection of small deviations from normal behavior in large amounts of data (credit card usage, mortgage fraud detection, etc.).

In much the same way that the natural immune networks successfully and efficiently protect organisms from such dangerous "errors" and invaders, we hope that in the future, the immunocomputer could be equally successful in controlling systems, computer networks, and even work in the human body as an implant.

References

Adamatzky, A., *Identification of Cellular Automata* (Taylor & Francis, London, 1994).

Adamatzky, A., Universal computation in excitable media: the 2^+medium, *Advanced Materials for Optics and Electronics*, 1997, **7**, pp. 263–272.

Ader, R., Felten, D. and Cohen N. (eds.), *Psycho-neuro-immunology* (Academic Press, New York, 2000).

Ageyev, V.S., Parasitic contacts of rodents in the river valleys of the desert zone of Kazakhstan and their significance in the plague epizootology. *Ph.D. Thesis in Biology* (Saratov, 1975, in Russian).

Agnati, L.F., *The Human Brain in Science and Culture* (Casa Editrice Ambrociana, Milano, 1998, in Italian).

Agnati, L.F., Fuxe, K., Nicholson, C., and Sykova S. (eds.), Volume Transmission Revisted. *Progress in Brain Research*, vol. 125 (Elsevier Science B.V., Amsterdam, 2000).

Aikimbayev, A.M. et al., *Epidemiological Plague Surveillance in the Ural-Emba and Ustyurt Autonomous Foci* (Gylym, Almaty, 1994, in Russian).

Alberts, B. et al., *Molecular Biology of the Cell* (Garland Pub, New York, 1986).

Alefeld, G. and Herzberger, J., *Introduction to Interval Computations* (Academic Press, New York, 1983).

Ashimov, A.A. and Sokolova, S.P., *Introduction to the Theory of Automatized Control Systems with Variable Configurations* (Gylym, Almaty, 1993, in Russian).

Ashimov, A.A., Ayaganov, Ye.T., and Sokolova S.P., *Automatized Control Systems with Variable Configurations for Objects with Delay* (Gylym, Almaty, 1995, in Russian).

Aubakirov, S.A. et al., *Instruction on Landscape-Epizootic Regionalization of Natural Plague Foci in Central Asia and Kazakhstan* (Almaty, 1990, in Russian).

Banzhaf, W. and Reeves, C. (eds.), *Foundation of Genetic Algorithms* (Morgan Kaufmann Publishers, San Francisco, 1999).

Batuev, A.C. and Babmindra, V.P., Module organization of cortex, *Biophysics*, **38** (2), 1993, pp. 351–355 (in Russian).

Bay, S. D., *The UCI KDD Archive* [http://kdd.ics.uci.edu] (University of California, Dept. of Information and Computer Science, Irvine, CA, 1999).

Bender, W., Gruhl, D., Morimoto, N., and Lu, A., Techniques for data hiding, *IBM Systems J.*, **35**(3–4), 1996, pp. 313–336.

Bersini, H. and Varela, F., Hints for adaptive problem solving gleaned from immune networks, *Proc. of the 1st Workshop on Parallel Problem Solving from Nature*, 1990, pp. 343–354.

Birman, K., Shiper, A., and Stephenson, P., Lightweight causal and atomic group multicast, *ACM Transactions on Computer Systems*, **9**(3), 1991, pp. 272–314.

Bohinski, R., *Modern Concepts in Biochemistry* (Allyn & Bacon, Boston, 1987).

Braudes, R. and Zabele, S., *Requirements for Multicast Protocols* (Network Working Group, Request for Comments 1458, 1993).

C and K Systems, *Global Security Technology* (1997, http://www.cksys.com/).

Cantor, C. and Schimmel, P., *Biophysical Chemistry*. Part 1: The conformation of biological macromolecules (W.H. Freeman and Co., San Francisco, 1980a).

Cantor, C. and Schimmel, P, *Biophysical Chemistry*. Part 3: The behavior of biological macromolecules (W.H. Freeman and Co., San Francisco, 1980b).

Casanova, G., *Vector Algebra* (French University Press, Paris, 1976, in French).

Catalogue of health indicators (World Health Organization, Division of Health Situation and Trend Assessment, Geneva, 1996).

Cheng, J and Kricka, L.J. (eds.), *Biochip Technology* (Harwood Academic Publishers, Philadelphia, 2001).

Chua, L.O. and Yang, L., Cellular neural networks: theory, *IEEE Trans. Circuit Systems*, **35**, 1988, pp. 1257–1290.

Conception of transition of Russian Federation to sustainable development (Decree of the President of the Russian Federation **440**, Moscow, 1996, in Russian).

Coutinho, A., Immunology: the heritage of the past, *Letters of the Institute Pasteur of Paris*, **8**, 1994, pp. 26–29 (in French).

Coutinho, A., The network theory 21 years later, *Scand. J. Immunol.* **42**, 1995, pp. 3–8.

Dasgupta, D., Artificial Neural Networks and Artificial Immune Systems: Similarities and Differences, *Proc. of the IEEE Int. Conf. on Systems, Man and Cybernetics* (Orlando, USA, 1997).

Dasgupta, D. (ed.), *Artificial Immune Systems and Their Applications* (Springer, New York, 9).

Dasgupta, D. and Attoh-Okine, N., Immunity-based systems: a survey, *Proc. of the IEEE Int. Conf. on Systems, Man and Cybernetics* (Orlando, USA, 1997).

Dasgupta, D. and Forrest, S., Novelty detection in time series data using ideas from immunology, *Int. Conf. on Intelligent Systems* (Reno, USA, 1996).

de Boer, R., Segel, L. and Perelson, A., Pattern formation in one and two-dimensional shape space models of the immune system, *J. of Theoretical Biology* **155**, 1992 pp. 295–333.

de Castro, L.N. and Timmis, J. Artificial Immune Systems: A New Computational Intelligence Paradigm. (Springer, New York, 2002).

de Meer, J., Mahr, B., and Storp, S. (eds.), *Open Distributed Processing*, **II**(C-20), (Elsevier Science B.V., IFIP, North-Holland, 1994).

Edelman, J.M., Topobiology, *Scietific American*, 1989, pp. 44.

Ekins, R. and Chu, F.W., Microarrays: their origins and applications. *Trends in Biotechnology*, **17**, 1999, pp. 217–218.

Farmer, J.D., Packard, N.H., and Perelson, A.S., The immune system, adaptation and machine learning, *Physica D*, **22**, 1986, pp. 187–204.

Fauré, A., *Perception and Recognition of Forms* (Edi Tests, Paris, 1987, in French).

Forrest, S., Javornik, B., Smith, R., and Perelson, A., Using genetic algorithms to explore pattern recognition in the immune system, *Evolutionary Computation*, **1**(3), 1993, pp. 191–211.

Forrest, S., Perelson, A., Aleen, L., and Cherukuri, R., Self–nonself discrimination in a computer, *Proc. of IEEE Symposium on Research in Security and Privacy* (Oakland, USA, 1994) pp. 202–212.

Forrest, S., Hofmeyer, S., and Somayaji, A., Computer immunology, *Communication of the ACM*, **40**(10), 1997, pp. 88–96.

Frisken-Gibson, S.F., Using linked volumes to model object collisions, deformation, cutting, carving and joining, *IEEE Trans. on Visualization and Computer Graphics*, **5**, 1999, pp. 333–348.

Gantmakher, F.R., *Introduction to Matrix Theory* (Nauka, Moscow, 1988, in Russian).
George, F.H., *The Foundations of Cybernetics* (Gordon and Breach Science Publishers Ltd., London, 1977).
Gilbert, C. and Routen, T., Associative memory in an immune-based system, *Proc. of the 12th Nat. Conf. on Artificial Intelligence*, Seattle, USA, 1994, pp. 852–857.
Ginsburg, S., *The Mathematical Theory of Context-Free Languages* (McGraw-hill, New York, 1966).
Gobran, S. and Chiba, N., 3D surface cellular automata and their application, *J. Visualization and Computer Animation*, **10**, 1999, pp. 143–158.
Goldfarb, L., Transformation systems are a more economical and informative class description than the formal grammars, *Proc. 11th IARP Int. Conf. on Pattern Recognition*, The Hague, The Netherlands, 1992, Vol. 2, pp. 660–664.
Gorelick, D.O., Kuznetsov, V.I., and Khvorov, G.V. (eds.), *Ecological Atlas of Saint-Petersburg* (Monitoring, St.Petersburg:, 1992, in Russian).
Gorodetski, V.I. and Tarakanov, A.O., Genetic like model for high speed networks, *Proc. of the 1st Int. Workshop on High Speed Networks and Open Distributed Platforms*, Tschammer V. and Smirnov N., (eds.), St.Petersburg – Berlin, 1995, pp. 189–193.
Hainzl, S., Zoller, G., and Kurths, J., Similar power laws for foreshock and aftershock sequences in a spring-block model for earthquakes, *J. of Geophysical Research B: Solid Earth*, **104**, 1999, pp. 7243–7253.
Haykin, S., *Neural Networks: a Comprehensive Foundation* (Prentice Hall Inc., New York, 1999).
Health of Population and Environment (monthly inform. Bulletin, Federal Center of Sanitary Inspection # 5, 10, 12, Moscow, 1997, in Russian).
Hogg, T. and Huberman, B.A., Controlling smart matter, *Smart Materials and Structures*, **7**, 1998, pp. 1–14.
Hori, T. et al. The autonomic nervous system as a communication channel between the brain and the immune system. *Neuroimmunomodulation 2*, 1995, pp. 203–215
Horn, R. and Johnson, Ch., *Matrix Analysis* (Cambridge University Press, 1986).
Hunt, J. and Cooke, D., Learning using an artificial immune system, *J. of Network and Computer Applications*, **19**, 1996, pp. 189–212.
Ishida, Y., An immune network model and its applications to process diagnosis, *Systems and Computers in Japan*, **24**(6), 1993, pp. 38–45.
Ishiguru, A., Watanabe, Y., and Ychikawa, Y. Fault diagnosis of plant systems using immune networks, *Proc. of the IEEE Int. Conf. on Multisensor Fusion and Integration for Intelligent Systems*. Las Vegas, USA, 1994, pp. 34–42.
Jerne, N.K., The immune system, *Scientific American*, **229**(1), 1973, pp. 52–60.
Jerne, N.K., Toward a Network Theory of the Immune System, *Ann. Immunol.*, 125C, , Paris, 1974, pp. 373–389.
Johansson, K.H., Egerstedt, M., Lygeros, J., and Sastry S., On the regularization of Zeno hybrid automata, *Systems and Control Lett.*, **38**, 1999, pp. 141–150.
Kanal, L., On pattern, categories and alternate realities, *Pattern Recognition Letters*, IARP-92 Conference Special, **14**(3), 1993, pp. 241–255.
Kitron, U., Landscape ecology and epidemiology of vector-borne diseases: tools for spatial analysis. *J. of Medical Entomology*, **35**(4), 1998, pp. 435–445.
Korneva, E.A., About the international conference "Interaction between nervous and immune system and environment", *Physiological J.*, **82**(1), 1996, pp. 140–143 (in Russian).

Kryukov, V.I. (ed.), *Discussion on Neurocomputers* (Science Center of Biological Research of the Academy of Sciences of the USSR, Pushino, 1988, in Russian).

Kuich, W. and Salomaa, A., *Semirings, Automata, Languages* (Springer, Berlin, 1986).

Kuznetsov, V.I., Gubanov, A.F., Kuznetsov, V.V., Tarakanov, A.O., and Tchertov, O.G., Map of complex appraisal of environmental conditions in Kaliningrad. In: *Kaliningrad. Ecological atlas (11 maps)*, 1999 (in Russian and English).

Kuznetsov, V.I., Milyaev, V.B., and Tarakanov, A.O., *Mathematical Basis of Complex Ecological Evaluation* (St. Petersburg University Press, 1999).

Kuzminov, A.Yu., On dependence of oligopeptide structure from amino-acid sequence, *Biophysics*, **32**(2), 1987, pp. 206–209.

Lallement, G., *Semirings and Combinatorial Applications* (John Wiley & Sons, New York, 1979).

Lamport, L., Time, clocks and the ordering of events in a distributed system, *Comm. ACM*, **21**, 7, 1978, pp. 558–565.

Lavrus, A.S., *Security Systems* (Science and Technik, Kiev, 1996, in Russian).

Liu, J.-D., Ko, M.-T., and Chang, R.-C., A simple self-collision avoidance for cloth animation, *Comput. and Graphics*, **22**, 1998, pp. 117–128.

MacBeath, G. and Schreiber, S.L., Printing Proteins as Microarrays for High-Throughput Function Determination. *Science*, September **289**(5485): 2000, pp. 1760–1763.

Maselko, J., Self-organization as a new method for synthesizing smart and structured materials, *Materials Science and Engineering*, **4**, 1996, pp. 199–204.

Mekler, L.B. and Idlis, R.G., General stereochemical genetic code — the way to the biotechnology and to the universal medicine of the 21st century now. *Priroda*, **5**, 1993, pp. 28–70 (in Russian).

Olshansky, A.Yu., *Geometry of Determinative Relations in Groups* (Modern Algebra Series, Nauka, Moscow, 1989, in Russian).

Passive infra red detectors, 1996, [http://www.delta-projects.com].

Perelson, A., Immune network theory, *Immunological Reviews*, **10**, 1989, pp. 5–36.

Plaksin, D.Yu., Raev, M.B., and Gromakovskaja, E.T., The method of stereospecific assay and the method of the conjugate for stereospecifis assay. *The Russian Federation Patent 2089921*, 1997 (in Russian).

Protein chip challenges, *Signals*, [http://www.signalsmag.com], 2000.

Rao, C.R., *Linear Statistical Inference and Its Applications* (John Wiley & Sons, New York, 1967).

Rao Vemuri, V. (ed.), *Artificial Neural Networks: Concepts and Control Applications* (IEEE Computer Society Press, Los Alamos, CA, 1992).

Romanovsky, Y.M. et al., *Mathematical Biophysics* (Nauka, Moscow, 1984, in Russian).

Rose, S., *The Making of Memory: From Molecules to Mind* (Bantam Press, London, 1992).

Salomaaa, A., *Jewels of Formal Language Theory* (Computer Science Press, Rockville, 1981).

Segel, L.A. and Cohen, I.R. (eds.), *Design Principles for the Immune System and Other Distributed Autonomous Systems* (Oxford University Press, 2001).

Shaitan, K.V., Dynamics of electron-conformation transitions and new approaches to physical mechanisms of functioning of biomacromolecules, Biophysics, **39**(6), 1994, pp. 949–967.

Shary, S.P., Solving the linear interval tolerance problem, *Mathematics and Computers in Simulations*, **39**, 1995, pp. 53–85.

Skormin, V.A., Delgado-Frias, J.G., McGee, D.L., Giordano, J.V., Popyack, L.J., Gorodetski, V.I., and Tarakanov, A.O., BASIS: a biological approach to system information security, *Information Assurance in Computer Networks* (Gorodetsky V.I., Skormin V.A., and Popyack L.J. eds.., LNCS 2052, Springer-Verlag, Berlin, 2001, pp. 127–142).

Sokolova, S.P et al., *Intelligent Security Systems* (ed. A.O.Tarakanov, Almaty, Police Academy of Kazakhstan, 2000, in Russian).

Stayton, P.S. et al., Streptavidin–biotin binding energetics, *Biomol. Eng.*, **16**(1–4), 1999, pp. 39–44.

Swanson, P., Gelbart, R., Atlas, E., Yang, L., Grogan, T., Butler, W. Ackley, D., and Sheldon, E., A fully multiplexed CMOS biochip for DNA analysis. *Sensors and Actuators*, **B 64**, 2000, pp. 22–30.

Tannenbaum, A.S., *Computer networks* (Prentice Hall, 3rd Edition, 1996).

Tarakanov, A.O., Semantic models for synthesis space navigation software. *Proc. of the 2nd Workshop on the Problems of Applied Space Ballistics*, Moscow, USSR, 1986, pp. 84–88, 195–198 (in Russian).

Tarakanov, A.O., Matrix method for automated synthesis of programs, *High School Letters, Industry*, **31**(10), 1988, pp. 21–25 (in Russian).

Tarakanov, A.O., Optimization of a class of interorbit transfers by the theory of catastrophes. *Letters of Russian Academy of Sciences: Technical Cybernetics*, **2**, 1992, pp. 77–81 (in Russian).

Tarakanov, A.O., Mathematical models of information processing by biomolecules: formal peptide instead of formal neuron, Problems of Informatics, Russian Academy of Sciences, **1**, 1998, pp. 46–51 (in Russian).

Tarakanov, A.O., *Mathematical Models of Information Processing Based on the Results of Self-Assembly* (Thesis for Sci.D. degree in physics and mathematics, St. Petersburg 1999a, in Russian).

Tarakanov, A.O., Formal peptide as a basic agent of immune networks: from natural prototype to mathematical theory and applications. *Proc. of the 1st Int. workshop of Central and Eastern Europe on Multi-Agent Systems* (*CEEMAS'99*), St. Petersburg, Russia, 1999, pp. 281–292.

Tarakanov, A.O., Information security with formal immune networks, *Information Assurance in Computer Networks* (Gorodetsky, V.I., Skormin, V.A., and Popyack, L.J. eds., LNCS 2052, Springer, Berlin, 2001, pp. 115–126).

Tarakanov, A. and Adamatzky, A. Virtual Clothing in Hybrid Cellular Automata. Kybernetes, *Int. J. of Systems & Cybernetics,* **31** (7/8), 2002, pp. 1059–1072.

Tarakanov, A.O., Bakhtin, A.E., and Kvachev, S.V., An expert system for real-time interpretation of objects, *Proc. of the Int. Conf. on CAD/CAM, Robotics, and Factories of the Future* (Shaal, H. and Ponomaryov, V. eds., St. Petersburg, Russia, 1993, pp. 377–382).

Tarakanov, A. and Dasgupta, D., A formal model of an artificial immune system. *BioSystems*, **55**(1–3), 2000, pp. 151–158.

Tarakanov, A., Sokolova, S., Abramov, B., and Aikimbayev, A., Immunocomputing of the natural plague foci. *Proc. of the Genetic and Evolutionary Computation Conference* (*GECCO-2000*), *Workshop on Artificial Immune Systems*, Las Vegas, USA, 2000, pp. 38–41.

Tarakanov, A., Sokolova, S., Abramov, B. and Dubyansky, V., Complex evaluation of plague epizootic as a result of pattern recognition, *TAUAR* **3**, Almaty, 1999, pp. 51–54 (in Russian).

Tarakanov, A.O. and Tumanov, M.V., *Modern Mathematical Methods for Complex Evaluation of Health* (Yussupov, R.M. ed., Anatolia, St. Petersburg, 1998, in Russian).

Tarakanova, N. and Tarakanov, A., Detection of similarity in the dynamics of infectional diseases by a mathematical model of molecular recognition, *Proc. of the 2nd Int. Conf. "Ideas of L. Pasteur in the struggle against the infections"*, St. Petersburg, Russia 1998, p. 213, in Russian).

Tesniere, L., *Elements de Syntaxe Structurale* (Klincksieck, Paris, 1965, in French).

Thayse, A. et al., *Logical Approach to Artificial Intelligence: From Classical Logic to Logical Programming* (Bordas, Paris, 1988, in French).

Thom, R., Topology and linguistics, *Results in Mathematical Sciences*, **30**, 1(181), 1975, pp. 199–221 (in Russian).

Toffoli, T., Programmable matter methods, *Future Generation Computer Systems*, **16**, 1998, pp. 187–201.

Toffoli, T. and Margolus, N., *Cellular Automata Machines* (MIT Press, Boston, 1987).

Toffoli, T. and Margolus, N., Programmable matter, *Physica D*, **47**, 1991, pp. 263–272.

Van der Waerden, B.L., *Mathematical Statistics* (Springer, New York, 1969).

Volino, P. and Thalmann, N.M., Developing simulation techniques for an interactive clothing system, *MIRALab Report*, 1998.

Wasserman, P., *Neural Computing: Theory and Practice* (Van Nostrand Reihold, New York, 1990).

Weimar, J., *Simulation with Cellular Automata* (Logos, Berlin, 1998).

Wilkinson, J., *Algebraic Eigenvalue Problem* (Clarindon Press, Oxford, 1965).

Witkin, A., *Physically Based Modeling: Principles and Practice of Particle Systems Dynamics* (TR Robotic Institute, Carnegie Mellon University, 1997).

Yang, T. and Yang, L.-B., Fuzzy cellular neural network: a new paradigm for image processing, *Int. J. Circuit Theor. Appl.*, **25**, 1997, pp. 469–481.

Zhu, Z.J. and Liu, C., Micromachining process simulation using a continuous cellular automata method, *J. Microelectromechanical Systems*, **9**, 2000, pp. 252–261.

Index

A

AB-network, 25
allosteric effect, 6, 18, 19
amino acids, 4, 5
antibodies, 7
antigen, 7, 20, 22, 78
artificial immune systems (AIS), 3, 8, 9, 10, 11, 130, 132, 148
artificial intelligence (AI), vi, 1, 2, 3, 4, 5, 9, 11, 41, 52
artificial neural networks (ANN), v, 2, 3, 9, 10, 11, 24, 41, 145
artificial neuron, 2, 11, 153
Atomic Broadcasting Protocol Based on Lamport Time Stamps (ABCAST LT), 92, 95, 96
attributive CF grammar, 79
attributive formal grammar, 23

B

BB-network, 24, 25, 26
B-cells, 4, 22, 23, 24, 26, 145
binding, 6, 8, 11, 17, 18, 19, 20, 22, 23, 54, 59, 147, 150, 175
binding energy, vi, 17, 18, 19, 24, 31
biochip controller, 177
biochip matrix, 175
biochips, 12, 147, 153, 173, 174, 175, 176, 177, 178, 181
biological specificity, 6
biomolecular computers, 12
biomolecular computing, 4
biomolecular interaction, 12
biomolecules, 175
bit string, 3

C

case (or role) grammars, 81
Causal Broadcasting Protocol Based on Vector Time Stamps (CBCAST VTS), 92, 96, 97
cell adhesion, 7
cell proliferation, 22
cell recognition, 7
cells, 2, 4, 9, 22, 108, 146, 148, 155, 164, 165, 168
cellular automata (CA), 2, 3, 11, 24, 100, 101, 145, 146, 147, 153
cellular immune networks, 101
cellular neural networks, 101
code indicator of the word, 62
compactness, 52
computer viruses, 10
condition of compactness, 58
context-free grammar, 76, 154
C-reactive protein, 180
cryptography, 164
crystallization, 6
cybernetics, 1

D

deflation method, 35
deletion, 63
diabetic nephropathy, 180
double helix, 4, 5

E

eco-information system, 126
ecological atlas of of Kaliningrad, 120

ecological atlas of St. Petersburg, 117, 118
evolutionary computations, 10, 11

F

folding, 5, 6, 11, 18
folding a vector to a matrix, 47, 49, 50, 56
formal allergy, 27
formal B-cell, 22, 23
formal grammars, 62
formal immune memory, 27
formal immune network (FIN), vi, 11, 22, 23, 24, 27, 100, 145, 153
formal immune response, 27
formal immunodeficiency, 27
formal logics, 62
formal protein (FP), vi, vii, 11, 13, 15, 16, 17, 18, 19, 20, 21, 22, 23, 59, 62, 81, 83, 89, 91, 98, 134, 145
formal T-cell, 22, 23, 154
FP-probes, 55
free energy, 5, 13, 16, 17, 18
free FP, 23, 24
free monoid, 61

G

genetic algorithms (GA), 2, 3
genetic code, 4
genetic memory, 7
genome, 7

H

homologous binding, 42, 45, 46

I

idiotypic networks, 7, 11
immune memory, 7, 27, 153, 183
immune networks, v, 6, 7, 8, 10, 11, 22, 97, 145, 148, 184, 185
immune system, v, vi, 3, 4, 7, 8, 9, 28, 61, 145, 146, 148, 153, 154, 163, 173, 182, 183

immunochip, vi, vii, 12, 145, 146, 147, 148, 149, 153, 154, 155, 156, 160, 161, 162, 163, 164, 179, 181
immunochip emulator, 165
immunocomputer, v, vii, 11, 145
immunocomputing (IC), v, vi, vii, 11, 12, 13, 41, 50, 55, 61, 78, 107, 110, 112, 137, 145
immunoglobulines, 12
immunology, v, vi, 8, 11
infection morbidities, 123
inherent grammar, 82
integer valued FIN (IFIN), 24, 25
intelligent security systems, 137

K

key and lock hypothesis, 6
knowledge-based reasoning, 78

L

Lamport time stamps, 95, 97
learning classifier systems, 9
learning sample, 55
Lighthill, J., 1
linguistic binding, 81, 82
logical clock, 96
logical time, 95
logical time stamp, 96
loops, 93, 94
lymphocyte clones, 12
lymphocytes, 7

M

matrix eigenlanguages, 65
microalbuminuria, 180, 181
microarrays, 147
mirror matrix, 85, 86, 87
molecular circuits, 6
molecular recognition, 5, 6, 41, 54
monopeptides, 18, 91
morphology, 63, 81
mutation indicator, 22, 23

N

negative-selection algorithm, 9, 10
network of binding, 22, 23, 24
neural networks, 1, 4, 6, 10, 11, 186
neurochips, 2
neurocomputers, 2, 10, 11, 12, 41, 146, 153, 183
neurons, 4, 6, 11, 146
nigrescence, 179

O

open distributed processing (ODP), 92
orbital hodograph, 113, 165, 173

P

parametric uncertainty of complex processes, 36
partial matching rule, 9
passive infrared detectors (PIR), 138, 140
pattern recognition, vi, 12, 41, 50, 52, 53, 54, 55, 143, 155
peptide spectrum of a word, 64
plague foci in Kazakhstan, 127
prefix, 73
prefix eigenlanguage, 77
processing, 64
proteins, vi, vii, 4, 5, 6, 9, 11, 13, 14, 15, 17, 18, 155, 173, 174, 175, 176

R

Ramachandran diagram, 168
recognition, 6, 7, 8, 9, 10, 17, 22, 44, 46, 61, 120, 133, 134, 161

S

SA/Bi technology, 177
scalar clock, 99
scalar time, 92
scalar time grammar, 99
scalar time stamps, 95

secondary structures of proteins, 87, 89
self-assembly, 5, 6, 16, 17, 18, 167, 168
semantic valences, 82
semiring, 65
separability, 52
shape space, 136, 144
singular value decomposition (SVD), 31, 55
specificity of recognition, 47, 49, 50, 53, 55, 56
stationary states of FP, 17
steganography, 163
structures of native proteins, 83
suffix, 73
suffix eigenlanguage, 77
supervised learning, 50, 55
surface CA, 111
surveillance of the plague, 130
synchronization protocols, 92

T

T-cells, 4, 22, 23, 145, 146
T-FIN, 98, 99
theory of linguistic valence, 81
three-dimensional FIN (3DFIN), 26
threshold of binding, 18
time based multicast protocols, 92
topology, 25
torsion matrix, 84, 85
two-dimensional FIN (2DFIN), 25, 170

U

unsupervised learning, 50, 55

V

valence of a verb, 82
vector clock, 97, 99
vector time, 92
vector time grammar, 99
vector time stamps, 96
virtual clothing, 100, 101